Optimizing Project Work, Management, and Delivery

T0272070

T033128

Gary L. Richardson and Brad M. Jackson

CRC Press
Taylor & Francis Group
Boca Raton London New York

CRC Press is an imprint of the
Taylor & Francis Group, an **informa** business

AN AUERBACH BOOK

First edition published 2024
by CRC Press
2385 NW Executive Center Drive, Suite 320, Boca Raton, FL 33431

and by CRC Press
4 Park Square, Milton Park, Abingdon, Oxon, OX14 4RN
CRC Press is an imprint of Taylor & Francis Group, LLC

ISBN: 978-1-032-49822-5 (hbk)
ISBN: 978-1-032-49366-4 (pbk)
ISBN: 978-1-003-43109-1 (ebk)

DOI: 10.1201/9781003431091

Typeset in Garamond
by Deanta Global Publishing Services, Chennai, India

Optimizing Project Work, Management, and Delivery

Thousands of project management–related books have been written. Why is *Optimizing Project Work, Management, and Delivery* different?

This book represents the authors' experiences gained from looking at the problem of project management for 50 years and wondering why projects cannot be more successful. Experience from various management models and techniques has helped but still does not fit reality or provide accurate forecasts. Industry surveys have compiled the root causes of project failure, and yet they persist. Is there no answer to this problem?

As the book explains, the management solution is not in the models or the theory but is found in how they are mapped against the actual target project characteristics. This is the book's unique strength. There are major coverage gaps in current project management models that also need to be recognized. All of the existing models are correct in some ways, and yet each is also wrong.

The book starts by reviewing popular models and related topics that help construct the building blocks of an integrated model structure, which is at the core of this book. The integrated model described here is meant to be a decision-oriented view related to the project life cycle rather than a cookbook of success steps. Project management is too complex for a cookbook approach. This text helps managers find that right path.

Contents

SECTION II   DELIVERY STRATEGIES

Preface

The journey to produce this book is difficult to accurately retrace because it was not what one would call logical. It all started with weekly Zoom discussions between the two authors who were cloistered away because of COVID. This was little more than two geeks who had been working in the project industry for decades talking about the past. During our professional careers, we both struggled with various projects in our different work environments. In 2018, we collaborated on writing an extensive overview of this topic titled *Project Management Theory and Practice*. As the weekly Zoom topics continued, we began to get more specific regarding failure-related items and what might work to mitigate these items. In general, this phase is best described as philosophizing about what is wrong with the current world of project management. At some point along the process, fragmented solution ideas began to emerge. The first one was how to embed multiple work execution strategies in a single project structure. That seemed like a band-aid to satisfy both the waterfall and agile crowds. As time progressed item after item was added to the picture as we began to define what the full management domain should include. The one conclusion that we saw at this point was a firm belief that projects were more complex than the current models supported and some of the defined problems were acerbated by typical band-aids used to cover gaps in those models. At this point, we had a lot of material and didn't know what to do with it.

As the weekly discussion continued, one of the first conclusions we agreed on was that none of the current project management models covered the whole life cycle and too much industry conversation was focused on trying to defend one model over another. As a result of this, we started looking deeper into various classic models and examined how they viewed a project. This helped to define gaps and design drivers in major models. Industry surveys helped to summarize key success/failure factors and the existence of a recurring theme of common failure factors. The generally poor deliverable success rates indicated that project managers do not seem to grasp these points or else they are not controllable. The research findings regarding time-phased project success by Dr. Ryan Nelson opened up yet another gap area to ponder. As the weekly discussions continued, we finally put sufficient puzzle pieces into a form that seemed to fit the problems identified. The design goal was to show a model that could be used for all project types and support multiple work execution strategies within the same structure. The very creative title for the new model was an "integrated model" (and there you have our level of humor). This collaboration process took approximately one year, but we had each been personally involved in the project world for more than 50 years and watched the management process evolve through many silver-bullet solutions over that period. Not knowing what to do with the draft model, we approached John Wyzalek, a technical editor at Taylor & Francis (CRC Press) who reacted favorably to the topic, and that eventually led to the product you are reading here.

Then came a reality! No more friendly Zoom sessions. Putting this material into the public eye became scary from an ego standpoint. In previous publications, the subject area had already been legitimized by another source so all required was to try to explain it better. In this case, the topic is highly controversial and the solution even more so. Management views on this topic are now being presented to technical groups with hard-core beliefs that their model is the best. This text is saying that they are all wrong to a degree, but all correct as well. Hopefully, that weaseling view will be sufficient to abort any assassination plots. At any rate, we do recognize that some of the statements made regarding some existing process will likely be disagreed with by the sponsoring camp. That's all right! Our hope is that this description of a very complex management process will be received as a legitimate analysis of the process from a source that was not trying to sell something.

The integrated model described here is meant to be a decision-oriented view related to the project life cycle and certainly not a cookbook of success steps. This topic is too complex for that approach, and the operative answer for a specific project has to come from enlightened management who knows how to look into the dark and find the right path. More details related to the evolution of this material are covered in Chapter 17.

We hope that the reader will accept this description as an honest venture into a fuzzy technical world and a reasonable technical analysis of the management process. To keep the size of this treatise into something less than a *War and Peace* novel, there were known shortcuts made. Specifically, process factors regarding the role of team management are left out but this is recognized as a major success factor. We ducked this by just saying that the team is motivated and highly skilled. That is another book for another day in a land far away!

Within the model description, there are multiple significant process modifications recommended to improve project outcomes. All of these are included to help cover identified gaps or practices in current models. The following six items represent significantly modified views in comparison to traditional management models:

1. Explicitly defining the target project's characteristics and goals as management drivers.
2. Recognizing the variability of project success definition based on target goals.
3. Recognizing the overall schedule damage caused by task padding.

4. Describing how multiple work types can be embedded within the same plan structure.
5. Describing the myth of traditional status tracking.
6. Questioning the role and value of top management as a project control layer.

We hope that this exploration of a revised project management process will improve your ability to structure a project for success and better understand what issues to deal with during the life cycle, regardless of which model you currently favor. Even more, we hope that you see the logic in this integrated view over any other traditional model. Believing that one of the popular existing management models is the best method and adequate for your project type may well represent a mind block that limits an improved outcome.

Gary L. Richardson

Brad M. Jackson

About the Author

Gary L. Richardson retired from the University of Houston, College of Technology graduate project management program as the PMI Houston Endowed Professor. During this 16-year academic period, he taught PMP and project-related programs for the university, plus external U.S. and international organizations. He has taught various project management programs to audiences in Finland, Russia, Iraq, China, and South America. He previously held professional certifications as a Professional Engineer (PE), Project Management Professional (PMP), and certification in Earned Value. In this time frame, he produced six professional texts on the topic of project management.

During the early phase of his career, Gary served as an officer in the U.S. Air Force leaving as a Captain. Following the military period, he held positions as a manufacturing engineer, at Texas Instruments, Consultant to the Comptroller at the Defense Communications Agency, Department of Labor, and the U.S. Air Force (Pentagon) in Washington, D.C. Interspersed through these positions, he was a tenured professor at Texas A&M and the University of South Florida, including adjunct stints at the University of Houston and Sam Houston State University. Following this, he moved to Texaco in various IT-related senior management roles, then finishing his industry career at a Texaco/Aramco joint venture and Service Corporation International where he held CIO-level management positions. In 1991 he was a finalist for Outstanding IT Executive for South Texas Region. During the early academic period, Gary published four technical IT-related texts and numerous technical articles.

Through this broad array of experiences, Gary was involved with over 100 formal projects of various types, which collectively provided a real-world laboratory of experience matched with an equally broad organizational function view. Throughout his career, he has observed management and technical issues frequently encountered and along with this has been an active participant in the evolution of project management techniques that have occurred over this time.

Gary earned his B.S. in Mechanical Engineering from Louisiana Tech, an AFIT post-graduate program in Meteorology at the University of Texas, an M.S. in Engineering Management from the University of Alaska, and a PhD in Business

Administration from the University of North Texas. He now lives in Houston, Texas, and can be reached at richagl0001@gmail.com.

Brad M. Jackson is co-founder/CEO of cordin8, LLC. He has spent over 35 years focused on developing and implementing collaborative technologies in support of team-based organizations. He has implemented these technologies to support teamwork for global clients in oil and gas, insurance, consulting, and telecommunications.

He began his career at Texaco where he led a team to explore the use of group decision support systems and other collaborative technologies to support teams as part of the Total Quality Management initiative at Texaco. That work included partnering with behavioral scientists at the University of Minnesota, including a jointly sponsored Texaco/NSF grant to study the impacts of group decision support systems on the TQM teams.

By the mid-1990s, he formulated a vision for an organization supported by collaborative technology in his paper "The Dynamic, Re-configurable Organization of 2025: Mastering the Interplay Between Information Technology and Organizational Design." It served as the backdrop for the blueprint of an organizational operating system, *cordin8*, that became the platform for the company that he co-founded, cordin8.

Prologue

Why Should I Read Another Project Management Text?

There have been literally thousands of project management-related books written. Why is this one different? This text represents experiences gained from looking at the problem for 50 years and wondering why projects can't be more successful. Experience from various management models and techniques helped some but still do not fit reality or provide accurate forecasts. Industry surveys have compiled the root causes of project failure and yet they persist. Is there no answer to this problem? This text will attempt to convince you that the management solution is not in the models or the theory. Rather, it is more related to how those are mapped against the actual target project characteristics. None of the current models looks at the problem this way. In addition, there are major coverage gaps in the current models that also need to be recognized. All of the existing models are correct in some ways and yet each is also wrong.

The first 12 chapters will wade through various background topics that help construct the building blocks for a new integrated model structure that deals with the observed gap issues.

Project Management Myths

This list of beliefs is at least present in a significant number of project managers.

- Agile principles represent the future direction for all projects.
- The waterfall model structure is the best model for all mature organizations.
- PMI's PMBOK (project theory guide) has the answer to project management problems.
- Project success is measured by schedule, budget, and functionality.
- Projects would be more successful if the approved scope was frozen for execution.

- Having padded estimates for each task is the best way to achieve on-time completion of projects.
- Carefully planning the project scope before execution is mandatory for effective control.
- Planned versus actual measurement is the key to good status tracking.
- Management must have a good forecast of the proposed project deliverables before approval.

If you answered positively to any of the questions above, there will be statements made in the text that you will disagree with. The author's challenge is to try to show why these statements are at least suspect and generally wrong.

Search for the Holy Grail

Personal Involvement in the project world began almost 60 years ago as a young engineer trying to build a high-tech widget, and then a few years later struggling to write a Ph.D. dissertation on the topic of project controls for the DoD procurement environment. Many other varied project experiences with both authors continued throughout our careers across broad industry segments. Those years of struggle with this topic led to the belief structure outlined in this text.

To level set the reader, it will be necessary to wade through some historical background. One must understand that the current forms of project management did not come down the mountain on a stone tablet written by some all-seeing guru. Quite the opposite. The history of this process has been somewhat chaotic and continues that way now. In many ways, it resembles a horse built with giraffe spare parts. As the key historical pieces are described in the first 12 chapters, you will see details of these fragmented and disorganized characteristics. Theses overview description may not be agreed to by all readers, but the hope is that there will be some insights that are unique and workable. Personal bias toward some particular model approach may make it difficult to change one's perspective on a new approach. That is the nature of complex topics.

The art of project management is much more complex than the average person understands, and many project managers exhibit behavior that supports this statement. Project failure rates remain high, yet various surveys have traced the root causes back to similar sources year after year. There is no easy answer to explain why these same sources continue to repeat. The text will attempt to show systemic processes' shortcomings for various common approaches and how one might go about dealing with the major ones. Space limitations made it necessary to eliminate trying to add human behavioral issues to this problem. To get around this, the text will assume the human members involved are motivated and knowledgeable regarding the current management approaches, but that is a suspect assumption.

Also, no motivational factors will be mentioned even though it is understood that there is much to say on this topic as well.

Based on this logic, the text will focus on a description of the revised management model structure and its elated key processes. The biggest challenge is to provide adequate evidence that there is a problem. This point suggests that the early part of the text must be dedicated to outlining problem areas. Otherwise, any new model presented would be rejected without first covering an assessment of key model gaps and then using that background to propose a new view. The solution outlined here contains decision layers and related best practice processes that are designed to minimize failure factors. Some of the items described will be familiar to many readers, but others will be new and more obscure. The integrated model described later in the text has the following characteristics:

1. It can be used in all project types.
2. It morphs itself around its target project characteristics and delivery goals.
3. It allows a lean approach by allowing variable degrees of process formality.
4. It allows both traditional and iteration work delivery strategies within the same project structure.
5. It supports a variable-defined definition of project success.
6. It looks at project scope as a graded variable rather than a fixed single fixed target.

Where to Look for My Keys?

There is an old joke that sets the stage for this text. It goes something like this:

As I left the project office late one night, I saw this guy under the street light on his hands and knees searching for something. I went over and asked him if he needed some help (he was obviously inebriated).

> *He said in his slurring response, "I looss my keesys."*
> *I said, "Where did you lose them?"*
> *He pointed and said, "Up the street!"*
> *I said, "Why are you looking down here?"*
> *His response was like thunder "There is no light up there."*

Maybe this scenario answers many project questions regarding why we cannot adequately find methods that significantly improve project success. Could it be that the industry professionals are looking in the lighted areas, but the answer is still in the darkside someplace?

So how can a strange story like this relate to project management? Over the years, there have been many notable theories, tools, processes, and other types of views to improve this process. Collectively, there may be so much that we are confused by the conflicting volume. An astrophysicist recently gave a talk explaining the Big Bang Theory and how one could now see into the distant past (13 billion light years ago). This dynamic look back in time was supposed to explain the situation today and estimate how long the universe would exist before it blew up. It was nice to know that we had a few more billion years but it is still not completely clear exactly what he said. In some analogous way, this is similar to the story of project management. Maybe the basic problem is much like the drunk guy looking for his keys and using Big Bang notation to hide the flawed logic. The project management process has been trying to construct tools and techniques to help guide the process, but the design premise may be wrong since so much of the looking has been in the wrong place. Maybe the current views have valid pieces but do not cover the whole problem but rather fragments of processes that offer an easier-to-understand view. The thesis of this text is that much of what has been derived thus far has been looking at the problem as a disjointed fragment of segments. Several project delivery models have been derived over the past 70 years, all subtly claiming to have contained magic delivery steps to achieve success. Each of these looked like it had merit but none have truly solved the management problem. In other words, the past models implied that if one followed these steps the project would succeed. These efforts did bring insights into some aspects of the overall problem but thus far have not offered great success when evaluated across the full project domain. Also, several interesting new non-dictionary terms were coined through this process so project professionals can talk more like astrophysicists. On the positive side, there has been increased recognition over the past few years that something was not right with the project delivery approach. Recognizing that some change is needed is a good mental starting point for improvement.

A starting place for describing a solution to this situation involves explaining the factors that are causing the problem. Also, it is important to define what role this text has in this mix. Simply stated, this text aims to describe a project management decision model that can be used for all project types. It will be driven by the characteristics of the target project and not by a fixed set of cookbook steps.

Design Philosophy

The following statement represents the core model philosophy design view for this text:

> *A project is charged with producing some defined output, therefore the process to achieve this goal should be to first evaluate the project characteristics, delivery goals, and from that examination define the operational environment. The management process should be overlaid on this view.*

Questions about the effectiveness used for managing current projects were the initial drivers that led to this text. Research into the current models validated the notion that they did not approach the project in the right direction. The classic design view seemed to have the belief that project data could just be poured into the model and the desired output would emerge but there was ample evidence that this was not the case. The unexpected conclusion from this research was a need to invert the traditional model's fixed process and focus away from a static type structure to one more linked to the specific target profile. We believe this is the key that leads to the proper structure and process. Also, it supports a better match for identifying the proper decision logic through the life cycle.

As with all such ventures, the result shown here may well be controversial in various areas depending upon the reader's personal background. The hope is that you will find merit in the review regardless of your bias on this topic.

PROJECT ENVIRONMENT

This section outlines selected background issues to level set the reader's knowledge regarding various aspects of this topic. The six chapters in this section are:

Chapter 1 Introduction—Provides some introductory background on the state of project management and its current practices.

Chapter 2 Delivery Methods—Introduces the concept of work delivery and the basic evolution of this topic since WWII. The current approach to delivery is shown to be ad hoc and fragmentary.

Chapter 3 Project Profiles—One of the key tenants of the text is that projects have different characteristics and delivery goals. This chapter describes how a project can be profiled.

Chapter 4 Evolution of Project Management—An overview of the key evolutionary steps is an important element in describing gaps in the current development models.

Chapter 5 Project Success Drivers—The industry survey offers quantification regarding the success rates of a project and the factors that influence those results. This chapter summarizes the major items that influence project outcomes.

Chapter 6 Project Externals—Externals as used in this chapter related to two factors that are external to the initial project plan. These are scope change and risk events. It is important to understand that they are external factors that emerge during execution to change the direction of the project.

DOI: 10.1201/9781003431091-1

Chapter 1

Introduction

Goal of the Text

The aggressive goal of the text is to describe an integrated decision architecture for project management that will fit all project types. This also includes key processes and tools that can improve the probability of project success by guiding key decisions through the life cycle for all project types.

After pursuing project targets across a wide array of types, we have concluded that projects are more complex than our models support and current techniques adequately deal with. Our research led to the conclusion that finding a better management method is going to take a deeper conceptual dive to improve deliverable success. Based on this belief, the writing goal of this text is to explore a more integrated decision structure of techniques to fill in the current gaps and from this produce a more coherent view for managing projects of all types. Given the complexity of this problem, this proposed solution is not amenable to a fixed cookbook of steps, but rather a hybrid view that will need customization based on the project type. In other words, the method will need to fit the project rather than having the project poured into a fixed model.

To present a new solution to this problem it is first necessary to describe various gap issues in current approaches. The new model is meant to focus on design logic more than detailed mechanics since the first step in accepting new approaches is to believe in the design concept. Also, this is not an attempt to destroy all of the current approaches but to show what elements of them work best and what bad practices most lead to failure. The text will evolve the pieces of this integrated model and include a summary of "success recipes" that need to be included in the decision process.

DOI: 10.1201/9781003431091-2

Reader Audience

The audience for this text is assumed to be mixed regarding background and technique bias. One group will approach the topic from their current bias which is most likely either traditional (waterfall) or contemporary (agile-centric). Both of these groups will likely be inclined to feel that their chosen approach is the correct one and this will take clear evidence to change that view. The second assumed audience is senior managers who want to understand more about why projects are not successful. This audience represents the ones who want to know in advance how much the project will cost and how long to complete prior to their approval. Any significant changes related to the new process would need to be supported by this group. A third group is the project team which is charged with delivering the target item. There are significant changes outlined in the model that affect this group. It will take careful analysis of this problem and clear logic for the new approach to convince these three groups that this is a better method.

The attitude we seek for the reader is to be objective concerning project management methods and concepts. It is important to recognize that all of the current models have some logic gaps compared to the real-world environment. For this reason, all current approaches can be improved. It is critical to have this view of the material and not just try to defend the current view. There is no attempt to call anyone's baby ugly, but rather suggesting that improving their homework would improve their chances for a better future. There are a lot of great ideas floating in the project industry currently, but essentially none of them look at the whole picture across all project types.

Author's Background

The authors of this text have been involved with the project world for decades (60 and 40 years, respectively). In reviewing this experience, it now feels like a *Search for the Holy Grail*, or maybe more accurately like the frustration of not finding *Noah's Ark*. In the early days, the targets were focused on defining tools for design, analysis, and status reporting (i.e., WBS, Gantt charts, schedule networks, Earned Value, DFDs, software tools, etc.). Later efforts focused on creating the magic cookbook of life cycle management steps—large books of documentation that no one seemed to read. As this random walk process moved into the 21st century, we suddenly recognized that hordes of others were pursuing the same golden dream and many claimed to have found it. Each of the guru's ideas was good in many cases but none seemed to solve the basic problem of producing successful projects. This assessment statement may be controversial as well as various parts of this text may also be.

Some quantitative studies by learned organizations quote reasonable outcomes for defined project types—those that use process X, or those that follow a certain management method. These may be valid statements but do not reflect the broad industry statistical surveys. Related to this point, we will explore the concept of success in more detail in the text but our experience suggests that most projects are not managed well and the results frequently do not fit planned outcomes. If the reader believes this assessment is not true, then the text is probably not for your reading pleasure.

Project Environment

Current project managers are exposed to innumerable management models, theories, and other strategies to make project execution more successful. Some describe these as *silver bullets* popularized years ago by the old Lone Ranger comics. Each silver bullet was advertised to kill all the bad guys and bring peace (just like the project manager is supposed to do).

The hybrid management process and related techniques described here will require a more analytical project manager who understands how to deal with the unique characteristics of their specific project. Also, there is recognition that the host organization's culture will be affected by this type of change. There must be recognition that the current models have significant gaps with reality. Organizations are fundamentally bureaucratic and fixed in their view of project work processes, making radical change difficult. Any new approach will have to show operational success in this domain before it can change the existing culture.

Current Practice

One of the complex issues related to this area is the requirement to produce a schedule and budget forecast before approval, and then try to deal with changes later in the process that upset even an accurate forecast. There is comfort in having a fixed forecast that defines the completion date as June 4 and a budget of $800,000, but experience says that neither of these parameters will be correct in the end. One must ask if it better to have a comfortable wrong answer or a less comfortable domain answer. Also, is it better to produce higher customer satisfaction or meet the forecast parameters? In essence, these questions represent the tip of the iceberg facing the project team.

There is a great debate in current project management practice, arguing about techniques to deliver successful projects. Some of the most popular models are named waterfall, Lean, Extreme, Scrum, and agile, all illuding that they are the best. The thesis of this text is that all of these are correct, and all of them are also wrong! They are correct when used properly when executing a specific project type but wrong otherwise. None of them provide the correct set of processes for every project, which will be further discussed later in the text. A proper project delivery strategy should apply model steps and techniques that fit specific project profiles and delivery goals. Also, the use of these model techniques described should improve the probability of delivery success. Many current projects are managed using some popular delivery model which is based on various assumptions that do not fit the project profile and this mismatch situation is often not properly considered. A common trait is to favor one known method and force the project into that format. The approach outlined here is to first evaluate the project's characteristics and delivery goals, and then from this design an appropriate management structure that fits the profile. By attempting to force the management process into a fixed model structure, it becomes necessary to band-aid gaps in the model to fit various situations. This approach often creates problems as bad as not having a model to work with. Some common band-aids seem to be used without this understanding. The thesis posed here is to suggest that the proper answer to this question will have to be a more flexible approach operated by better-trained managers who understand how to use a more flexible delivery model, rather than a fixed structure that doesn't fit. Second, it needs to be recognized there is a vast collection of current technical management documentation and experience surveys that collectively describe the processes that need to be followed to achieve project success, yet it is found that the same issues continue to be repeated over time. For example, if experience says that projects fail most frequently because of user and management support, why does that lesson have to be relearned? There are so many practical examples of this that it makes one wonder if the project team was recently imported from another planet and has never seen a project before. In other situations, it is obvious that the project management culture does not understand that they have a customer while they blindly plow forward with an "optimum" task sequence based on their internal view. Models work best when they match reality and that is not the normal case with the current set of options.

The Need for Speed

The action movie *Top Gun* has popularized the term "the need for speed" and that same theme is relevant in the project world. Indeed, projects do not deliver value until they deliver actual results, so delivery speed is a universal need, but the answer to this is not always to move as fast as possible. Throw up an office building as quickly as possible and then having it fall violates the basic speed requirement. Likewise, producing the wrong widget or business process violates the value equation and is equally bad. Any singular goal statement about a process can be wrong for these types of reasons. The use of projects to create new business products or processes is much more complex than the average participant realizes. If that were not true, why is it common to often find 50% of the projects undertaken result in failure? This is a significant waste of organizational resources that must be tagged as a management responsibility. Can all projects be made successful? Likely not!

But can they be made more successful with a better management approach? Most assuredly yes! We do hope that the processes and prescriptions outlined in this text will make all projects more successful. Following this architectural approach should produce a noticeable improvement in matching delivery goals. But this is not an automatic process. It requires technical decisions to be made as described. Organizational cultures have to change as well and management expertise is still needed. We believe that a more enlightened approach to the management process as described here will result in more successful outcomes.

Background Theory

This text will not go into a myriad of details regarding all the things that need to be done in managing a project. That ship has sailed. The Project Management Institute and other professional organizations have done this sort of broad description already and these sources are considered remedial background reading if one wants to work in this profession. One must take from this ocean view, extract a set of theories that can be used in all project types, and from this select the level of formality that fits the specific project requirement. The use of an integrated decision and process structure as outlined here will help guide a more coherent delivery approach. This view represents the underlying theory of the text.

Model Design Concepts

To effectively use this material, the reader must accept the notion that all projects are different and they have different delivery goals. No one model best fits this scenario. From exposure to this background material and design logic, the reader should come away with a broadened view of project delivery techniques and some of the key success processes that will improve outcomes. There is extensive literature regarding sources of project failure and the highest probability from that list, however, each project has to be viewed based on its unique characteristics rather than a survey list. The material described will attempt to weave this sort of thinking into the new decision structure. The general idea of that view is if one avoids failure, there is an improved chance of success. Unfortunately, one can avoid all of the failure factors and still not achieve success.

The integrated model will take usable pieces from various models and fit them together into a "Snap-On" structure much like Lego blocks. As a prelude to showing this, it seems necessary to build the solution logic case by presenting background material that shows why the current models do not fit reality very well and outline some related "gap" issues that make these models fall short. From this background overview, a list of both best and key success-oriented practices emerges. Many of the items described are not necessarily new ideas but are often misused practices in the current culture that affect positive outcomes. In some cases, the process described is a misused theory that results in poor outcomes. Another aspect of this review process is to show how that model's base assumptions may not fit particular project types. Throughout this, there will be a mention that some portion of a particular model might have more general use if it could be embedded into another delivery structure. This is the Lego block idea. At the end of the classic model reviews, the goal is to identify a collection of worthwhile delivery techniques that have value in customizing project work.

Much of the current prescriptive literature regarding project management falls under the category of a 1930s vaudeville act where the doctor is being asked to fix a problem. The skit goes like this:

> *The patient says, "Doctor, doctor my arm hurts!"*
> *The doctor replies, "When does it hurt?"*
> *The patient says, "When I do this!"*
> *The sage doctor then says, "Don't do that!"*

A significant portion of project management advice follows this "don't do that" theme. Frequent advice is to motivate the team, define the requirements, estimate the task, etc. Yes, that is good advice and we are not rejecting those items here. But there is still the question of how to manage all of this and get to the mechanics mentioned. Some of the specific prescriptive tools imply that if you perform items A, B, and C, a desirable outcome will result but they neglect to remind you that the internal assumptions that led to this prescription do not fit your project. Project managers and senior leadership participants are going to have to become more knowledgeable regarding the proper management strategies for a project if they hope to achieve better results. It has always been an interesting phenomenon in a sadistic sort of way that the role of a published formal project plan is to provide all with a specific task list, completion date, and budget. Never mind that none of the items on this have a chance in hell of being correct. Ask yourself this question. Is it better to publish a discrete value for these parameters that are wrong by definition, or would some other approach that more accurately reflects the situation provide better communication? In other words, would a more general description relay a better view? Would a probability distribution showing range estimates provide a better forecast for completion than a discrete value? The answer to these questions seems pretty obvious if you think about it. This is an example of organizational culture forcing a bad management practice. Things of this sort are examples of the subtle psychological or political side of this topic. Changing the cultural side of this equation will be more difficult than publishing a better delivery strategy that is not consistent with that environment. The complexity of organizational change may well lie at the heart of this problem.

Historical Overview

The 70-year history of modern project management evolution has introduced several very creative ideas regarding how to manage this phenomenon. Interestingly, the most recognized outcomes of this have been simplistic solutions that often guide the process to an erroneous answer. These approaches have satisfied managers who don't want to think about added complexity, but future project managers are going to have to do a better job of educating their bosses rather than following overly simplistic demands from that source. It is one thing to require a fixed forecast over which reality does not exist, but a more enlightened idea is to understand how to manage the factors that produce the desired outcomes. Gantt charts remain the most common planning artifact seen in the project world. They are so simple that anyone can read them and interpret the defined status. The fact that the project will not follow this schedule does not seem to bother the audience. If the chart is produced by simply showing bars on a schedule, this may not be appropriate for millions of reasons. Software tools such as Microsoft Project may take a network project task plan and "magically" reformat that into a Gantt-looking chart, but even this modern mechanical transformation is not enough. That view is still static and does not truly represent the potential additions that are going to change the view. More examples of this miscommunication could be shown here but the important idea is to recognize that the project complexity makes all forecasting more error prone than the discrete measures that are so common.

Chapter Logic

Deciding on a text topic sequence outline was more difficult than initially thought, and the selected answer switched multiple times. The key topic sequencing dilemma was how to explain the result without defining the problem or solution pieces first. Would it be best to first show a schematic of a theoretical answer, and then dwell later on explaining various core components that would be used in that model? Or is it best to summarize the issues to resolve and some of the solution pieces, and then explain an integrated schematic? This second alternative bottom-up approach was the chosen option. Early chapters describe various pieces of the puzzle and summarize some of the factors that correlate with project failure. For example, what are the success management issues with risk and scope change? Second, what are the key assumptions underlying the most pursued models in use today? This peeling of the underlying problem onion consumes the first several chapters before attempting to start mixing and integrating key pieces. The primary difference in this approach versus the "doctor, doctor my arm hurts" model is that key success-oriented tidbits are uncovered from the best practices view. By using this approach, high-value "management nuggets" can be defined for use in designing a more robust project delivery model that has the flexibility to fit all project types.

Some readers will have previous knowledge in one or more of these familiar model areas, but few readers will be knowledgeable in all that is covered. Hopefully, this reader-leveling exercise will provide the necessary logic to substantiate the resulting model description and key success factors. To understand the integrated model, it is necessary to have a level reader background, particularly in understanding the mechanics related to variable work management logic and intermixing work types within the same structure.

One of the areas omitted from the text is the human team aspect. This is a text scope constraint and not because that topic is irrelevant to success. There is extensive literature on this topic, and we did not feel that this could be adequately mixed into the core structure-oriented theme. The suggestion on this topic is to review all of the current literature and then spend your life getting better at it. Most project managers feel like they can study the topic of human behavior for a lifetime and still have questions.

Much of the early evolutionary period was focused on the development of a management tool kit and basic techniques to improve isolated processes such as requirements definition, plan development, and status analysis, (i.e., items like WBS, Gantt charts, schedule networks, earned value, DFDs, software tools, etc.). Later efforts focused on creating the magic cookbook of life cycle steps accompanied by large books of documentation that no one had time to read even though they seemed to describe everything one would want to know about how to do the process. As this historic random walk process moved into the 21st century, we suddenly recognized that hordes of others were pursuing the same golden dream and many claimed to have found it. Each of the guru's ideas was good in many cases, but all seemed to be niche ideas and none seemed to solve the basic problem of producing successful projects. Much of what we describe here may not fit a particular reader's view of the problem and may also be a controversial conclusion. In many cases, we will not credit a published source for an opinion. As stated in the text, from our experience we have witnessed this opinion from multiple sources, and now state it here as our opinion. As we review our experience matched to all of the published sources, the conclusion reached is that the answer to managing a project is not adequately visible in the published models, but may be somewhat successfully practiced in the way the work required matches the model used. We are also sensitive to the failure rates of projects. If your car only started 50% of the time, that would not be considered acceptable yet this is the world of projects. One possible conclusion to that statistic could be that the audience believes the outcome of these ventures cannot be improved. We don't believe that is true. Better improvement in project results lies in the hands of the management process and its associated organizational culture.

As mentioned earlier, there have been many *silver-bullet* solutions to this problem over the years. The project audience grabbed many of these new solutions, like lemmings jumping off of the proverbial cliff, and supported each with almost religious fervor. Each of these seemed to bring some better understanding of the process but still produced marginal improvement. We will look at some of the historical evolution of key management practices over the post-WWII period, much of it sponsored by the U.S. Department of Defense (DoD) in large, high-technology product-type projects. Even though these projects produced some very interesting products, their overall delivery scorecard for schedules and budget is not impressive, yet the majority of all projects continue to follow this same model with minimal variation. The commercial project industry exhibits similar negative results. There is an old story about moving the chairs around on the Titanic and expecting a different result. That seems to apply here.

We now know a lot about what goes wrong on projects and the key areas involved in executing the process. The challenge here is to convince you, the reader, that there is an approach that if followed will improve outcomes. One can study this subject for a lifetime and still not be able to guide a successful effort. Some projects have failure built into their design. However, continuing to not look critically at the issues underlying this process will almost assuredly stagnate these trends. Our goal here is to help lay out the key mechanisms and tools that can successfully drive the project life cycle. As we now begin to dive into this complex subject, be prepared to not agree with everything stated and that is okay. Just try to view the subject from a broad perspective.

Chapter 2

Delivery Methods

Introduction

The approach used in this text to describe various project management processes is non-traditional in the sense that does not describe the full details of each model. Instead, the approach used is to summarize a broad view of the target process and show what management gaps need to be filled and then lay out some of the elements that a more integrated real-world method should include. The key at this stage is to look for existing work components and techniques that can be used to fill the observed gaps. To do this, it is necessary to describe key issues that exist with current methods and highlight how these gaps impact project success. There are many different models in the literature that purport to explain a better way to manage a project. Each of these is based on an assumption set and each of these may do a reasonable job of guiding one through a project that fits this unique assumption set. However, any one of them could be a bad choice for a project that does not fit that set of assumptions for a different target project type or delivery goal. There are multiple broad assumption tracks found in the various models described and each of them is narrow in its view of the environment. In each case, these models specify a set of processes to achieve the desired outcome. Most of the models exhibit both structure and process views. What is recognized at this point is the broad and unique characteristics of projects who have quite different management needs and output goals. When one attempts to use any management model it should match the characteristics and goals of its target since the model design is to aid in achieving some inherent goal. In the modern project management culture, there is not a clear or explicit process for ensuring that the existing project fits a particular model, nor are the output goals for the project explicitly defined. One gets the feeling that the implicit goal is to "do this and good things happen." Failure to deal with this shortcoming is one of the major failure-oriented aspects of the modern process. In

DOI: 10.1201/9781003431091-3

addition to this, many users do not understand why inserting the wrong project into a mismatched management structure is wrong. It is not logical to say that if the model is good, the result is also good. Based on this belief, the text may be critical of a particular model characteristic but that does not mean the model is bad, just a questionable fit to the target goal. All of the models described here have interesting characteristics, but it should be recognized that they also have characteristics that may not fit a particular situation. It may even be reasonable to conclude that none of them fit when viewed in the broader critical light of a specific project type versus a target set of design assumptions.

It often seems that a local project management approach is based on one model as though all projects fit that model. Seldom are the underlying assumptions of the target model even reviewed in comparison to the current venture. One of the themes that will be mentioned numerous times is the proper approach to managing a project must recognize the needs of the project and the management tasks associated with it. For example, is enough known about the delivery requirements to specify in detail all of the tasks required to deliver the proper goal? Would it be best to carve out a subset of the project that had shaky requirements, and then let the project team elaborate the output with active user support? How do you measure ongoing status for that type of situation? These types of project variability needs occur all the time, yet the use of a fixed model often hides those needs in pursuit of some model goal. There will be project managers who would reject this idea by saying that they would alter the plan accordingly. If so, that is a band-aid and such efforts can significantly modify the internal integrity of the model and distort what it is intended to accomplish. In this case, the model is not of great help. Management flexibility decisions of this type are legitimate and the model should simply reflect them and manage accordingly. All of this suggests that the structure of the model must have the flexibility that matches that need.

There are many confusing aspects of this mismatched target problem and finding a simple answer is no small challenge. A starting place to look for solutions is to explore the design logic of various key models used in the industry. Each of the ones selected has a good reputation among some segments of the project community. Some are touted as doing a good job of executing project work quickly, producing higher user satisfaction, or offering good status measures. All of these are valid points but each of these can be refuted when faced with a set of project profile goals that do not match that view. These design gaps are relevant and any one of the models can be shown to have marginal value in the wrong situation. The basic theory of a management model is that it aids in the underlying decision process but only if its design is focused on the right structure and assumptions. The proper use of such models is to apply the right processes that aid in producing the desired output—i.e., speed of completion, control, define requirements, etc. Is speed the high-level goal, or is it user satisfaction? Is there a significant risk in the venture? What about a situation where key resources are the constraint? Most current project environments recognize that work unit time estimates are not accurate, so they add "padding" to the estimates in the hopes that this will accurately reflect the completion time. This is an example of a process band-aid that exacerbates project overruns. This is not specifically a model gap but a cultural gap that creates negative behavior factors that are not understood. Decisions of this type cause projects to get off course because the model structure is not appropriate for the underlying process and the result is a poor outcome for the project. A better picture of these issues is needed before we attempt to outline more of this side of the equation. Hopefully, these examples have stimulated sufficient interest in moving forward through the underlying complexity. The follow-on goal is to seek out pieces of management techniques that can be re-assembled into a more appropriate integrated management view that properly deals with a broader array of project types and goals. Selling this approach to the technical organization groups is ambitious so it will take clear evidence that some significant change is justified without destroying the current views. In support of this, there is some visible evidence in current project literature recognizing gaps in the existing management process. Unfortunately, much of that seems to be arguing that one existing model is better than another, rather than trying to outline what the fundamental issues are for all models.

There are many cases where a major project failed for reasons other than the way it was produced. In other words, the target was produced as defined and yet failed in operation. If the goal is to achieve success, the management process must explicitly define what that is and design the project work effort around that. Success is not a simple variable or fixed for all projects. As a somewhat extreme example of this, Boeing designed and built a great 737-900 MAX airplane that suddenly crashed and caused the grounding of the fleet for an extended time. What management failure caused this to happen? Similarly, the Fukushima nuclear plant in Japan was operating successfully until a tidal wave wiped it out. Mother Nature and human behavior are cruel factors in the project world. These two negative events could have been avoided if the airplane had been properly tested for the design error, or if a valid risk assessment had identified the location threat from a tidal wave. If one is going to design a process for success, it needs to include decision components that lead to success. In the two examples above, there were both testing and risk process utilized so just having correct processes does not necessarily equate to success, but not having them certainly adds to the potential for failure. Project failure can occur from both poor execution of a process and its albescence. There are many more examples of a failure element not being included in the work plan. It would be instructive for the reader to browse the public post-accident reviews for the NASA Challenger explosion to see how the failure occurred even with sound project management. We only offer these negative points to highlight that this environment is much more complex than most understand. These events all occurred in a mature project environment. A management process must include not only the right steps but a culture that believes in the process and understands how to execute it. A model is more than a management pill to solve the problem.

Maybe a starting position statement for this discussion is to say that no single model for project management will be an appropriate answer in all cases, maybe even in any case. Projects have subtle differences that need to be dealt with in formulation, execution, or control. The basic challenge here is how to unpeel this complex onion and produce an understandable solution.

Delivery Strategies

From an abstract point of view, the goal of project management is to:

> *use appropriate skills and processes to complete a series of technical steps to deliver the defined goal. The operational process of project management is intended to execute the necessary work in an optimal manner which has the highest potential to accomplish the defined output.*

A search for techniques and models to aid in executing project deliverables has been a visible documentation activity since WWII (the mid-1940s). Tracing the historical evolution of this management process is somewhat difficult since there has been no central sponsor's guiding hand and many of the items developed are disjointed. Through much of the early period, the management items creation process took on the flavor of defining a project management tool kit to support project scheduling and budget processes. Some items developed along the way could be considered evolutionary, while others have more of a revolutionary bent. Collectively, most of the contributions focused on designing techniques to help define and control the various evolving life cycle processes and then map these to recognized topic areas such as scope definition, schedule, budget, risk, communication, and other life cycle processes. Early management definitional efforts were extracted from experiences on large governmental high technology, tangible product ventures (e.g., airplanes, missiles, ships). From these efforts, a traditional project view became engrained in the culture of organizations and even personal life. Simply stated, if you want to get something done, create a project. Even cleaning the family garage is now viewed as a project. Organizations increasingly recognize the project term as an important activity in producing new products and processes to achieve strategic competitive goals. There is likely no term in the business world more used that the term "project."

Blind Men and the Elephant

As stated earlier, the evolution of project management theory has not unfolded in an orderly process. In an attempt to find a memory parable to explain this phenomenon, we ran across the story of the *blind men and the elephant*. This story relates to the 19th-century poet John Saxe's parable of the six blind men who were asked to describe an elephant. This hypothetical situation seems to fit our story. Here is a sample of what the blind men found:

1. The elephant's leg feels like a tree
2. The elephant's ear is like a fan
3. The elephant's tail is like a rope
   Etc.

Each of these blind men was correct in their evaluation of an elephant, but in the end, none knew what an elephant looked like or what it could do. In this metaphorical way, this is much like the world of project management. There are numerous valid pieces to what we know but our project elephant is still a little mysterious in terms of what it looks like and how to tame it.

Current Perspective

The audience for this text is assumed to be mixed regarding their background involvement in the project role, technical background, or current bias. One reader subset will view the topic from their current experience which is assumed to be either traditional (waterfall) or contemporary (agile-centric). Another group may be interested in why projects don't succeed, and there is certainly insight here into that topic. Some readers will likely be biased toward their chosen management approach as being the correct view. Yet another potential audience is senior managers who want to understand more about why projects are not successful and what could be done to improve the process. A warning for this group is that you could be the root cause of the problem as much as some potential technical guiding mechanical approach. Unfortunately, the existing bias of this higher-level management group includes those who require information about the project before technical requirements are known and then later want to compare what happened to the early plan that now does not represent the ongoing project. Each of these groups has a different perspective regarding the goal of their project and this further complicates the management problem for the project team.

The traditional approach to project development specifies that requirements are defined and a work plan produced. Experience reveals that much of this early planning requirement causes excessive administrative overhead to produce erroneous answers that are then tracked downstream to reflect that the project is overrunning the plan. This cultural problem represents one of the major management time wasters and the material shown here may make senior managers conclude that their oversight goal is not productive as currently implemented. If a process takes excessive time and does not deliver useful output, it should be questioned. Inaccurate output projections from poor planning data are one of the primary sources of this result. That goal may have to be modified in the interest of improving the overall result. For all of these various stakeholder groups, suggesting any significant process changes will need to be supported by understandable logic and that is a recognized goal for this material. To buy into a change in management approach, the project participants will need to understand what is wrong with the current approach. Obtaining project data is one thing, but having actual project-improved results would be even better. One of the subtle stakeholder goal conflicts is to recognize that it is not adequate to define project success by simply meeting the original plan parameters of product functionality, schedule, and budget. Significant changes approved through the life cycle generally destroy those parameters, even with correction values made. So, if the project tracking system now says the project is overdue based on the originally approved plan, should the project team be viewed negatively for this overrun when the changes were approved by a formal management process? Logic says no, but this is a common scenario. Defining project success is more complex than any of the current management models imply. In reality, it may be comforting to have a project plan that defines the completion date as June 4 and a budget of $800,000, but experience says that neither of these parameters will be correct in the end. So, is it better to have a comfortable wrong answer or a less comfortable domain answer that is technically a more accurate prediction? A host of such similar countervailing goals needs to be better understood before we can deal with this conundrum further.

Our hats should go off to an individual who can take a fuzzy project requirement with a fixed schedule and budget, and go through the decision steps necessary to organize the work required to produce that item successfully within time and cost constraints. This is the ability we seek for the reader here as we begin to weave through this management maze. One once described project management as "getting used to living in the asylum." Some days that is a funny story and others not so much! At the very least a project manager has to be good at stress management.

There are a lot of great management theories floating in the project industry currently but essentially none of them are looking at a complete life cycle and organizational picture mapped across all project types. In this venture, we are trying to take a macro view of alternative ways of examining the project life cycle. Current project managers are exposed to a broad array of project management models, theories, techniques, and other strategies, all touted to make project delivery more successful. These are often called *silver bullets*, meaning a new solution to some problem. After personally pursuing many of these solutions for many decades, we have concluded that projects are more complex than our silver bullet models and current techniques often do not fit the environment in which they are used. There is a fancy term for this called *verisimilitude*, which essentially means that the model or technique emulates its target and can therefore yield a better understanding of the target and help to produce more accurate results. Various future chapter descriptions related to selected management processes and models will show that this is not the state of current models. If a model does not have verisimilitude, one way to resolve that is to produce another model. This new effort will fix the old view by identifying various new ideas that resolve the old problem. Unfortunately, there are still other old problems remaining unresolved. A more interesting approach is to look at each solution as only one component of an overall solution. Most of the proposed fixes only selected one small part of the problem to deal with. One must first start looking for an answer by defining the broad management domain issue— where does a project start and how do you look at the various ways in which it can be structured? Projects can be viewed as part of a larger program group of projects or they can be structured into phases based on key requirements or functionality. A lower-level management view may look at the problem by major components or functions. All of these views are variable and based on the way the problem is viewed, but each of these represents important management steps in the overall process. At the execution level, there are multiple ways of looking at a collection of work units—fixed structure, loosely defined structure, high speed, consulting with stakeholders, limited resources, etc. The concept of managing explicitly defined variable work execution techniques is important in a proper model structure. This term is meant to imply that tasks can have different goals for execution based on various characteristics of the work. These are valid management issues to deal with and any new model needs to support this. This characteristic must be included in an integrated view.

All of the concepts outlined thus far suggest that a proper solution to this problem requires a more flexible design. It will also require a more analytical project manager who understands how to deal with the unique characteristics of specific project types. Assigning a project manager who only knows about the technology may well lose that flavor. A proper management process requires both the technical and management aspects. There are numerous project situational examples clearly illustrating that the project manager or host organizational culture does not understand many of the tenants of good management. A clear vision from these sources is a critical element in the solution. There is an old and continuing debate in this industry asking what is the most important attribute of a project manager? One who knows the underlying technology best or alternatively one who is an expert in the types of management issues described here. Having both traits sound like the best option but is seldom the reality. One way of wrestling with this question is to answer a companion question of which one can you best do without. If you have no internal knowledge related to appropriate management needed and simply have a technical leader, you will focus on the technical aspects of the solution. However, if you only have management knowledge and no clue how to technically deal with the product delivery requirements, then that answer is also not promising. But, having access to both skills in the team regardless of the position of the leader can be a workable answer given appropriate personalities. Plainly stated, both skills are needed for effective delivery.

Organizations tend to be sluggish in their speed of change and fundamentally bureaucratic in their approach to executing projects. As a contemporary example, the approval of the agile iterative approach to project development has taken twenty years to occur. Even here, the model is being looked at as another silver bullet and not being incorporated into other known valid existing processes that should not be ignored. The goal of this effort is to recognize innovative ideas but not lose sight of the bigger picture in the process. Any new approach will have to show through experience that it better delivers desired outcomes but it also has to be understandable in concept. Both the traditional waterfall and agile concepts are simple to initially understand, but as we will show both have shortcomings that need to be dealt with.

Observing actual projects represents good self-learning case studies regarding factors that can improve success. Through the years these real-world project observations have been the personal research labs that lead us to the views outlined in the text. One of the lab examples in the text will be observations made at a local long overdue construction project where nothing seems to be getting done and long queues of cars are stacked up in the construction zone. This will be used to show how the management effort is oblivious to the true project goal. In this case, the question is what should the local project manager do, and what work management concept would have shown that improved direction? As professional project managers, we could find many similar examples. The construction model fits the tangible product model (waterfall) pretty well but even in this well-matched situation, many different work management decisions could be done to improve that output even by just understanding where the model does not provide proper direction. In the construction example, the project team was "perfectly" executing the design approach but not dealing with the core project goal of removing congestion. This example is not meant to pick on one industry, but it is a visible target that offers good common-sense examples of marginal management practices if one expands their management view as this text is attempting to do. This example project lost its delivery goal completely and was doing little to minimize the overall impact of the overrun. Evaluating project success in this example offers some food for thought. Time to complete is a major consideration to the user, but we suspect the construction company is more concerned with cost control and maybe even has resource supply issues. The question here is what should be the mitigation strategy when the project is not going well. Other less visible projects would show the same unmatched goals. Here is another visible lab example. If a company wishes or claims to have customer service, should that mean more than a website with no linked access to humans, or have phone wait times of 30 or more minutes because they "were experiencing higher than average traffic?" These everyday examples are very educational for understanding what appropriate project goals should be and how they impact their stated goals.

After the initial background work was done to compile what seemed to be a good starting place to examine this situation, the next conundrum became how to explain a new model logic without burying the reader in detail. The approach selected here is to initially focus on process gap logic and an appropriate overall scope of the structure. Various classic tools offer pieces of the desired process and will be described as one of the Lego blocks in a broader integrated model. In general, the overall focus will be on "what" is to be done more than the specific "how to." The integrated model result is not looked at as the final technical user guide containing all the details necessary to use the model. Frankly, we don't see that as a shortcoming. Flexibility allows less specification and more understanding of the output goal.

Here is a final philosophical note regarding the chapter layout approach. Projects do have significant life cycle similarities but it is also important to recognize that they also have distinct differences that dictate a flexible management approach. In the design view described here, the integrated model will morph around a defined target project profile and delivery goal set. To do that it must have decision steps that deal with these related variable needs. With this as a somewhat philosophical introductory overview, we are now ready to start unpeeling our project onion layers.

Chapter 3

Project Profiles

Introduction

This chapter deals with techniques to quantify differences in project delivery targets based on the unique characteristics and delivery priorities related to that target. The coined vocabulary term for this is a project *profile*. Projects can be described by categorizing delivery requirement characteristics and other related variables. Each of these defined variables affects the way a project should be managed. From this overall profile view, different work management techniques should then be matched to the defined priorities. Using this approach, one should see that each project's life cycle is unique and so too is the best way to manage that venture. The concept of managing the project based on its unique set of characteristics is one of the core beliefs of this text approach and is not explicitly dealt with in the current models.

Real-World Project Categories

In 2003, Russell Archibald published his research on defining a catalog of project types. He identified ten major categories and an 11th unnamed group for new entries (Archibald, 2003, 5). The ten major defined project groups are:

1. Aerospace/Defense
2. Business organization change
3. Communication systems
4. Formal Event projects
5. Facilities projects
6. Information Systems
7. International Development

DOI: 10.1201/9781003431091-4

8. Media and entertainment
9. Product and service
10. Research and Development

Within the ten major groups are 36 more specific types. Examples of these are a broad landscape such as movie production, energy systems, African aid, and the development of a new high-technology product. One of the best categories to illustrate a different management development strategy is a research and development project. In this goal type, it is often difficult to specifically define the outcome and the work process is often multiple steps of trial and error. In this case, the work process may be best designed using iterative techniques with multiple attempts to generate the desired output. This class of project is somewhat unique in its work management approach and control aspects. It is also interesting to point out that this type of project may produce something quite different from the envisioned outcome (e.g., a new drug, or new product). Not only did the project not produce the original goal but ultimate success might come from another alternative undefined deliverable. Many other examples illustrate how complex the proper design of the project life cycle definition can be. Modern movies are a visible class of project that often costs over $100 million with initial incomplete fuzzy scripts during planning and execution. Collectively, this broad breadth of various project forms sheds insights into project variability related to both the need for different management approaches and a flexible design of the underlying task characteristics.

Project Profiles

There are many ways to catalog a project's characteristics. Some obvious classification parameters are:

- Ability to define specific deliverable requirements
- Size measures—dollars, resources, time
- Previous experience with the project type
- Tangible physical product versus intangible product
- Skill level of the project team
- Ability to prototype the deliverables
- Level of management oversight required
- Availability of appropriate resources
- Level of perceived risk

To confound the project profile definitional issue even further, recognize that these parameters can occur in various combinations—i.e., large scope, new subject, tangible product, requirements not clear, and tight budget control required. The key point here is to describe how each project has a unique profile of these types of characteristics and this type of profile specification should guide the management process used. As an example, if a project is profiled as a small, familiar topic, clearly defined, and with a non-critical schedule or budget control, it would most properly be managed quite differently from the more complex specification outlined above.

Another important management guidance variable is the quantity, quality, and location of available resources. This factor is not a specific project deliverable characteristic but an environmental one that also affects the management style needed. The concept of variable project characteristics is not a new idea but it is often not visibly considered within fixed structure models. The recommendation here is to make this type of profile evaluation explicit in the planning process.

The question to deal with here is how to define the project profile and then use that as guidance subsequently to define the management approach for the related work. Every project manager intuitively realizes that projects are different but this is seldom explicitly dealt with in the life cycle management design of the project. Once the character of the project and its related deliverables is clearly understood, the customization concept becomes an easier idea to embrace.

Another key characteristic that impacts the appropriate management process is called *elaboration*. This vocabulary term relates to the notion of how project requirements are expected to evolve in clarity through the life cycle. One of the major management design factors related to this concept is the initial state of project requirements. If it is assumed that all technical and stakeholder resources can work together to specify the requirements at the initial planning stage, it will be possible to execute the project as a defined task effort. However, if requirements are not clearly defined at this point, the concept of a more iterative process of defining the required outputs becomes more likely as the preferred method of execution. The assumed decision on this point is one of the most controversial issues regarding how to best manage the effort. More on this topic is reserved for later sections of the text.

Management Delivery Strategies

Surveys of project performance over the years have indicated that the management and technical approaches used can affect the success of the outcome. In the early historical period, the delivery approach was somewhat ad hoc, but later this evolved toward a more fixed cookbook of tasks. Throughout the evolutionary period, theorists offered designs and tools for new management approaches that were touted to improve outcomes. Many of the historic models represented niche ideas suitable for a narrow spectrum of project types. Several examples of these will be summarized below. To simplify this view, we will catalog these efforts based on four underlying assumptions that drove those efforts. From this, three models are selected as the best surviving examples of those assumptions and the ones having the most promise for being used to combine into an integrated management view. Each of these historic attempts had some interesting design twists and this reveals some of how practitioners envisioned gaps in current practice. The breadth of these views is yet another example of the complexity of the appropriate underlying management process. Three of these evolutionary models have been selected for further expanded description here based on their fit for use in an integrated approach to the life cycle. Each of the selected models has characteristics that can be partially extracted and used in an overall approach and each deals best with some particular project profile characteristic.

A review of these development efforts viewed the management process as a set of evolving tasks to completion. All were essentially fixed in their view of the project assumptions and none explicitly started with an assumed flexible project profile definition. For example, the Critical Chain model (Chapter 10) focuses on techniques to speed up completion, while the agile/Scrum iterative model (Chapter 9) is touted as a method to improve customer satisfaction, among other characteristics. Finally, the traditional cascading fixed task predictive (Waterfall) style model (Chapter 8) assumes that project tasks can be defined during the initial planning phase. This model contains the most management bells and whistles based on years of maturation and use. Each of these classic models is best suited for specific model profiles, but none covers the desired breadth of recognized management domain or explicitly deals with the variability of profile characteristics—i.e., speed, customer satisfaction, risk, or control. Biased users with each of these models will likely disagree with this statement since they likely have defined band-aids to patch observed gaps. It is the gap question that we are focused on here. A technical description regarding how one might use different characteristics of these models within the same project is much more complicated than we are prepared to deal with at this point, but each model is worthy of further examination.

We have now outlined a description of real-world projects; it should not take additional proof to validate that projects can have different delivery goals and even constraints that can impact how they might be best managed. From this, it seems reasonable to postulate that a proper management process should involve these traits. The key remaining question is to define what tools or techniques might be applied to deal with various traits. Another less obvious management consideration here is to better understand the reasons for project failure. Those considerations also need to be embedded as management focus points.

During the early period, much of the published project management theory was based on experiences from highly complex governmental projects where the delivery focus was essentially on the defined functionality of the output device. Through later periods, topics such as schedule or budget began to emerge as associated major constraints for the project. In some cases, a project would be canceled because the cost went beyond the perceived value of the deliverable. In traditional project management, the "iron triangle" of three factors is often used to define success. These are functionality, schedule, and budget. Some sources are starting to add risk to this list with the logic that a project that fails because of some unanticipated reason is still not successful even if the other design parameters are achieved. In more current views, many of the historic models represented niche ideas suitable for a narrow spectrum of project types. Project success should no longer be viewed as a fixed set of parameters but more of a ranked priority of outcome parameters. For example, where does customer satisfaction fit in? In recognition of this evolving view of success, that aspect of the project also needs to be included in the definition of delivery goals. Here is an unusual example of how this can affect the process. Suppose a project deliverable is deemed so important that it must satisfy a defined customer at all costs. That single goal then becomes the main driver. Beyond that, the availability of required resources might be the next design factor. The point of these examples is to highlight how deliverable characteristics can affect the work management design process.

Methods to integrate various project work delivery methods are the goal here, rather than simply following a fixed model and assuming it is the singular correct answer for all cases. The following sections of the text will describe various selected "tidbits" of the overall optional definitional process and then describe an overall view that incorporates these. Related to this goal is the notion of reusing as many validated methods as possible, as well as using artifacts from previous projects to save planning time. The Lego analogy is used many times to emphasize that the proper operational pieces must fit (snap) together to serve their defined purpose. Each of the pieces will have a restricted role but collectively they will cover the full timescale of the project from birth to death. The first challenge in doing this is to identify a framework on which one can define the sub-pieces.

Each of the classic management models described in the text has evolved out of their unique project goal environments and each has a specific value in this environment. Notably, each of the approaches has a rigidly defined task work structure, and because of this, the management process would have to be patched to accommodate any other needs. One model might have an excellent structure to control but less value for customer satisfaction. A second model may make use of the assumption that the goal is vague and users are embedded in the process to actively review interim outputs. A third model might view the project as extremely time sensitive and shift all focus toward achieving that goal much like a track relay race with a baton—there is no fixed schedule, just go as fast as you can. As one can see, no one of these views fits all situations.

The analysis of project results over time has produced many lists describing sources of failure and some prescriptions regarding how to better achieve success. In many ways, there are mirror images of each as avoidance of failure seems to move toward success in most cases. Most of the models produced throughout this history have implied that they aid with success. This traditionally meant delivering according to stated requirements, on time, and on budget. That may be somewhat true of the project but what if the original goal of the project itself was wrong? The wrong project successfully delivered is still a failure. This is Just one more complex item to deal with in this puzzle.

Hopefully, the various examples outlined here have been sufficient to validate the point that there is more to delivering a project than following a fixed model, and that characterizes the state of the industry today. The total delivery life cycle must be part of the equation and the management of that process must understand the types of variables outlined here and be knowledgeable enough to know how to structure all of that into a delivery model. It is no longer appropriate to learn one of the classic models and attempt to force the task work into that structure, even if one tries to apply band-aids to that to make it fit. Recognition of this flawed management approach requires that future project managers become more knowledgeable concerning how different work management schemes impact the new definition of delivery requirements.

Defining Work Strategies

Once a project's deliverable priority goal set has been defined and approved, the next major question is how best to develop the specific task management approach. For instance, would it be best to set high-level requirements and let the project team and stakeholders work together to construct prototypes that evaluate the requirements, or should the deliverable be technically designed (blueprinted) from firm requirements and then produced by the team as designed? This requirements definition decision is one of the major guiding factors that will dictate the appropriate future management strategy. Experience shows that customers' changing views on requirements lie at the heart of many messy project management issues. In the traditional environment, a special process is added to the model assumption to handle this, called *scope management*. In some cases, this process alone is enough to destroy the integrity of the base model and in any case, creates unwanted management outcomes regarding forecast results and control. This is one of the most common band-aids added to the traditional model and occurs because of some level of variation in the initial assumption of scope definition. If a project experiences a 2% change in its scope every month, you have a much different project over one year, yet this often goes unrecognized. If the deliverables can be reasonably defined during the planning cycle, the required work can then be better work planned, results measured, and controlled. For example, if the design of a building can be specified through the use of standard "blueprints," a skilled technician can generally construct the desired building as specified. This is not to suggest that such a method is the best way to construct the building but it is a model that is used extensively. An alternative to this would be to loosen the design definition and let the project team decide how to execute some specific components. Here is a philosophical test of this concept. If you deliver a pile of miscellaneous materials, skip the planning stage, and start constructing a great playhouse for kids, do you think this is the best approach for dad satisfaction from the customers? Most would choose some level of planning, but how much? Recognize in this example that lack of a plan impacts the ability to accurately procure material or define what the future structure will look like. However, neither of these factors may be relevant. Loosening the original requirements definition might give the kid customers more input into what results and this may well be the premier success measure. We stated that projects had become ubiquitous and this is just one personal example. In our example, we only had one stack of material available so that was all they got (reduced satisfaction based on resource constraint).

There are many more examples and profile conditions to illustrate the management aspects, but the key points regarding the need for delivery flexibility seem sufficient proof. In some cases, it is good to remember the tortoise and the hare story from childhood—remember, the faster hare did not win. There is no single variable one can preset that will answer the question of how to manage all projects. If you come away with that understanding, this chapter has achieved its primary goal.

Profile Rating Codes

If a project is to have variable characteristics, the question becomes how to record that idea. Table 3.1 shows how the project's characteristics can be graded and used for future management decisions. The factors used here are examples and should be specifically defined for each project. The result of this effort is to be able to explicitly show priorities for management or delivery factors. This information would be used in both the design of the approach and during execution when some tradeoff in approach is required. As a simple example, assume that schedule is more important than functionality. This will indicate that efforts should be made to cut out some of

Table 3.1   Project Profile Parameters

Factor	Rating									
	1	2	3	4	5	6	7	8	9	10
Schedule										
Budget										
Functionality										
Resource issues										
Technology level										
Risk level										
Internal politics										
External politics										

the "nice to have" features if a cutback is needed to meet other goals. Many other tradeoffs can be made through this type of information.

Each of the profile areas needs to be rated on a scale of 1 to 10, with 1 meaning minimal concern and 10 meaning high concern. A brief explanation of each area follows:

Completion Schedule. All projects have timely completion as a goal. This factor is used to rate the criticality of the completion date.

Budget. The cost of a project is a key variable. This factor is used to rate the criticality of the budget.

Functionality. This factor is an important goal of the project. How much flexibility is there in this parameter set which can include multiple item grades?

Resource Issues. Project success is highly linked to the availability of adequate quantity and quality of project resources. This rating indicates the adequacy of this project.

Technology Level. Pursuing high levels of new technology has an adverse correlation with projects meeting planned delivery parameters. This rating reflects the degree of new technology related to the project goal.

Risk Level. Project risk levels can be influenced by both internal and external factors. This rating factor is an overall combined measure of perceived risk.

Internal Politics. This set of factors relates to various disagreements that may exist within the organization. This rating grade is meant to relate the impact of factors within the organization that creates confusion in the life cycle. Typical sources for this include conflicts from the departmental, managerial, project team, or other internal sources. A high rating here indicates the need for formal stakeholder communication processes.

External Politics. A wide variety of external stakeholders can create barriers to project success. This rating reflects the likelihood of such an impact.

Management Impact of Project Factors

Realize that the graded factors have a collective impact on the project and dictate certain management work priorities. The key point to understand here is that there is both a strategy trade-off role and a work management focus aspect to each. Concerning the rating factors shown here, the scores help guide management work design and tradeoffs. For example, if a project has a constrained budget limit, it would suggest extra management focus on the restricted resource. Likewise, a high-risk project would require more effort on the initial risk assessment and ongoing risk tracking. Low rating factors indicate a lesser focus priority on that factor. This approach to using project characteristics to guide management focus is intended to shape the project around those factors, rather than the current more static bureaucratic approach indicating that everything is critical and has to be pursued according to a fixed plan. One factor may be so dominant that it dictates a specific management approach. Traditional models do not offer an easy means to formally segregate such priorities, which then leaves the work design focused on all possible items. A project manager often has to make decisions in situations where a particular goal has to be lessened in favor of a higher-level one. These are called trade-off decisions. Note that some items on the rating list represent output deliverable issues, while others might represent design or internal constraint factors. All of these can influence how best to manage the project or view that variable during execution.

Rubber Boxes

Imagine the project and its associated aggregation of work tasks as being represented by a collection of rubber boxes. Each box has the dimensions of time, cost, and functionality. If all goes well, each box maintains these three defined dimensions. However, the real world often presents itself differently, and decisions are needed to be related to these various boxes. In the case of the overall project rubber box, the question becomes the relative importance of schedule, cost, and functionality of the final result. If it were possible to spend more resources to achieve the required functionality, the manager could increase that variable, which would also increase the budget dimension and hopefully improve the functionality of the project. This type of resource trade-off decision can be also used at the task level to influence desired outcomes. Note the analogy of the rubber box shows how trading off one of the variables to achieve change in one of the others. A second example is deciding to decrease functionality which could decrease the budget or schedule. The concept of "rubber" here is meant to say that these three variables can be adjusted by increasing or decreasing one variable to achieve a change in another. A more nonsensical case occurs when a senior manager suggests that it is possible to cut the budget and schedule while still keeping functionality constant. The deliverable rule of thumb is to be able to change one dimension and possibly affect another in the opposite direction. The concept of goal tradeoffs is an important consideration to understand. The only way to intelligently make this type of decision is to have goal priorities established as outlined here.

Project requirements can often be looked at as having must-haves, highly desirable, and nice-to-have levels. One industry term used for this is MoSCoW, which is an acronym for Must-have, Should-have, Could-have, and Want. If this graded requirement approach is used for requirements definition, the management planning goal might say that only the Must-have level is required. The remaining levels will be pursued as time and resources allow. This somewhat flexible view of scope as a "rubber box" is an important management idea that is often ignored. As project plans start to drift away from the approved level, the trade-off decision logic moves to the forefront of management practice.

Summary

This chapter has reviewed some high-level concepts related to structuring the project for execution. Current project management models have a static work structure that does not explicitly recognize this. As a response, band-aids are attached to the structure to deal with the gaps. The issue that clouds this bad practice is the degree of apparent commonality at the macro level across all project types. In other words, all projects are assumed to look alike. It is primarily at the lower level where major differences begin to be more significant. Industry practitioners have increasingly begun to "tweak" and expand traditional management views into new methods of planning and execution. Unfortunately, too many of these try to start completely fresh and ignore all of the past experiences and models. The thesis for this text is that there are many usable approaches in the current catalog if one can learn how to put the most usable pieces together in a meaningful way. This aggregation approach would simplify the learning curve involved and offer more evidence of validity. Most of the new models do not offer that much breadth of view at the work level as they have more static task-level perspectives and ignore the goal and profile issues outlined here. Based on this logic, the goal of the text is to show how this strategy can be structured into an improved way of looking at the management process.

Reference

Archibald, Russell and Vladimir Voropaev. 2003. Project Categories and Life Cycle Models: Report on the 2003 IPMA Global Survey, 18th IPMA Project Management World Congress, Budapest, June 18–21, 2004.

Chapter 4

Evolution of Project Management

The early project period from 1945 to 1960 was a period of developing management tools mostly for planning and control. By the 1960s, the project model had been recognized as an effective organizational model for developing complex products. Major management processes began to be formalized after 1960. Many of these were sponsored by the Department of Defense for use in their procurement practices. The typical target project for this period was for the development of large, tangible, and high-technology products (i.e., airplanes, ships, etc.). Major time and cost overruns became quantified and visible during this era. By the 1980s, formal management models were in place and various technical industry sources were beginning to become involved in documenting the management process. Focused schools of management process thought emerged. The waterfall model was predominant, but after 2000 there was increased attention focused on various iterative approaches. The management culture is not settled at this point.

In reviewing project management evolution from the 1940s to the present, one is struck by the ad hoc nature of how current practices and approaches have evolved. Some notable characteristics of this evolution are as follows:

1. The creation of projects seems to be more departmental in origin than strategic and competitive oriented.
2. Until recently, the majority of management approaches essentially followed a single life cycle view with minimal integration or crossover to new paradigms.
3. Project managers are often not trained in formal project management models or other academic aspects related to project management. In many cases, they are selected more based on experience related to the technical side of the

DOI: 10.1201/9781003431091-5

equation and not so much for their knowledge of formal management model theory.

4. The most common approach to managing projects has been through the waterfall model and its various defined work processes. Required project deliverables are assumed to result from these work units.

5. Most organizations would not be judged as having a mature support structure for projects.

As a result of these trends, the typical project management approach is characterized as immature regarding deliverable results and poor success rates. That said, even organizations that are described to be mature in this arena also have problems with repetitive success. This characteristic speaks to the subtle complexity of projects. One interesting phenomenon regarding project culture is observing the use of a similar process in two different projects and finding two completely different success outcomes. There is more to this topic than attempting to follow a list of required practices. Described another way, there is an implicit assumption that using some fixed prescription of processes and tasks will always achieve successful outcomes.

Environment Factors

There are many ways in which the current project environment can be described, but the following list represents frequent descriptions of environmental characteristics:

1. Before projects are approved, senior decision-makers require estimates of schedule and cost using high-level views of deliverable requirements.

2. There is often a management conflict dilemma related to project deliverable goals regarding time, cost, functionality, and resources. If achieved, the traditional theory says that meeting these criteria equates to customer satisfaction so long as the defined deliverables are produced.

3. During execution, shortcomings in pursuing the initially defined and approved project scope create change requests which increase schedule and cost, yet the original plan may remain fixed—i.e., more scope, but the schedule and cost plan remain fixed.

4. Task estimates for defined work are difficult to produce accurately, so estimates are padded to protect the plan, yet overruns remain for those tasks.

5. Projects contain varying degrees and sources of risk events that are hard to anticipate, yet these unknown events can significantly affect the project outcome. Techniques to manage this aspect of the project remain varied and are considered immature in accuracy.

6. Planned resources are often not available as defined or assumed by the formal plan, which often creates schedule and cost overruns—i.e., even a valid technical plan fails here.

7. Many projects use transient resources from multiple sources and skill groups which create a complex personnel management issue—i.e., training, morale, communication, and coordination are a few related issues here.

More examples of these difficult-to-handle environmental situations exist, but this sample list provides a reasonable starting point for the management issues to be dealt with.

Historical Path

To set the stage for describing improved management processes it will be a useful background to take a quick historical look at the path of key milestones that have led to the current view. Each of these major classic tools, techniques, or processes was created to deal with some perceived management aspects of the problem related to some deliverable need. In other words, the content of current popular models contains both assumed solutions and processes to deal with a particular management area (i.e., schedule, budget, risk, resources, scope change, etc.). As a result, the process of managing projects has evolved through a fragmented trial-and-error set of management models, tools, and processes. The section below briefly describes some of the more memorable project-related historical management model milestones.

Early Management Evolution to 1960

During the early 1900s, the management school led by Frederick Taylor developed various tools to improve manufacturing outputs. This culture of "scientific management" was focused on work process improvement. As a part of this movement, the Gilbreth's formal time estimating technique and Henry Gantt's famous 1907 graphical bar chart schedule work plan has remained visible in the management culture for over one hundred years.

Large product-oriented project experience from the WWII era created a formal awareness of the project's role in producing complex deliverables. Following this trend into the 1950s, the U.S. Department of Defense (DoD) began to document and prescribe various project-oriented tools and techniques. These efforts were often viewed as the first "silver bullet" solutions for project management (e.g., the answer to how to manage a project). One most notable example of process definition from this stage came from the very complex and successful U.S. Navy Polaris missile development project. Two strategic management tools were introduced in this project. First, a variable-time task estimating tool called the Program Evaluation and Review Technique (PERT) was introduced to define variable task estimates. Second, this project introduced a task network model for planning and defining the overall schedule and critical path. Both of these new approaches were credited with making this very complex project a success (that conclusion was later refuted by the way). Over time, the underlying arithmetic complexity related to the PERT algorithm calculations caused its usage to decline in favor of simple single-parameter task time estimates, even though this is now recognized as a backward trend in modeling project schedules. However, the critical path network concept for schedule planning remains an accepted planning and control tool typically known as CPM for Critical Path Management (or Model). Two other strategic tool techniques emerged in this time frame—Work Breakdown Structures (WBS) and Earned Value (EV). Mil Std 881 continues today as a definitional source standard for DoD's WBS family of products (i.e., airplanes, ships, tanks, etc.). Usage of the WBS schematic format is now a ubiquitous tool used across all industries, although there is no commercial movement to standardize this format. Meanwhile, EV usage as a status parameter is conflicted, and organizations are often not operationally mature enough to produce these parameters, even though there is evidence that this concept represents the most robust status-tracking tool available for the traditional management model. In recent years EV has been accepted as an international standard but still has weak industry sponsorship. Nevertheless, these four classic tools have been widely recognized as management thought leaders, even though only two of them are still actively used (WBS and task CPM networks).

By the end of this period, the concept of a project life cycle of tasks became the accepted view of projects. Another evolutionary thread over this time was the publication of formal project specification documentation by various organizations. Most of these "how-to" specifications were based on a predictive cascading task waterfall design model structure that is characterized by specifying the project scope and work definition before execution. From this specification, a formal schedule and budget can be produced before the project is started, even though experience shows this to be poorly correlated to the actual result. The concept of predefining project work and from that producing schedules and budgets remained a key concept in the evolution.

Phase Two Trends

Starting after WWII, the emergence of the Cold War brought an almost panicky period of increased levels of military spending for a wide array of new weapons. Figure 4.1 shows three high-technology airplanes (X-24A, X-24B, and X-70), which were three initiatives undertaken during the 1948 to 1966 period, none of which went into production. The X-70 project was eventually canceled with the claim that the cost was excessive for its planned function.

Numerous other less exotic projects were also being pursued during this period, and the government's goal was to find the proper method to control the schedule and cost of such projects. The management process emerging here was to produce a completion prediction for the schedule and budget from deliverable goals, then track those variables through the development process. This cultural approach to project management is visible today and the design roots of the classic waterfall

Figure 4.1 High-technology projects (1948–1966). Source: USAF Air and Space Museum

model are found in these historic projects. If one wanted to be critical of this early era, it would be in the strategic visioning side, as well as the immature approaches to planning and control.

By the 1960s, publicly available data from large DoD projects revealed for the first-time tremendous variances in planned time, budget, and functionality. From this initial recognition, more serious efforts were undertaken to define and formalize improved management and control models and processes. Five notable initiatives are selected here to highlight key "silver bullet" solutions for solving the project management problem. Since their inception, none of these have solved the overall problem of avoiding failed outcomes but many of these are part of the relics found in current methods. Recall the previous story of John Saxe's parable about the six blind me describing an elephant (Chapter 2). All of the silver bullets outlined here are somewhat correct, but none described what the project elephant looked like in total. Each of the various sponsored management specifications has attempted to solve the project delivery problem by defining "do all of these things and the result will be better." It is important to give credit to these efforts as each describes a potentially valid aspect regarding how to manage the project environment. Each of the five areas outlined below added insight to a formal management starting point. It is important to recognize that this evolutionary process spawned new insights into this complex process. Collectively, the following five epoch management specifications represent the basic thought ideas driving management since the 1960s. The date-grouped summarized list follows:

1. 1960s to present—various *DoD initiatives* to define and document project management
2. 1970s and 1980s—*Process definition tools.* Various manufacturing and software industry initiatives focused on speed and structure. Just-in-Time (JIT) delivery process techniques and structured programming represent this domain group.
3. The 1970s to present—*Professional methodologies.* This segment relates to the emergence of professional organizations that published prescriptive models to improve the management of projects.
4. 1950s to present—*Quality management programs.* In response to successful Japanese experiences, various programs were introduced to improve the customer satisfaction aspects of project initiatives. This improvement activity took on a general character of project management but more of a human bent than true process.
5. 1960s to present—organizations began to recognize that projects should be viewed similarly to stock investment by evaluating alternatives and making a more strategic selection. This school of thought became known as *project portfolio management* (PPM).

Each of these evolutionary threads left its mark on the current environment and for that reason, it is important to describe them here in a little more detail related to these trends.

DoD Initiatives

There are many examples of DoD documentation related to project management and much of this source material has migrated into the commercial environment through their contractor providers. One significant example, of a strategic document first introduced in the 1960s, is the recognition that a valid project architectural view for organizations is part of the required view of the overall management requirement. The initial specification document, DoD 5000.1, was titled the Cost Schedule Control Systems Criteria (CSCSC) and it defined 32 management "criteria" (processes) that government contractors had to meet as evidence of their project management control maturity. Later versions of this document began to focus more on Earned Value project status certification and the underlying specifications related to the original 32 key processes remained in view (DCMA). This document represents a classic definition of an organizational architectural view of project support control processes and can still be used to define that portion of organizational project support maturity.

Process Reengineering

The linkage between process reengineering and project management is not immediately obvious. However, if one views a project as a collection of linked tasks, that overall view begins to resemble a project life cycle process. From this perspective, it is possible to see how this formal process design movement began to influence project management practices.

Professional Management Models

One of the most popular international management models visible today is sponsored by the Project Management Institute (PMI). In 1986, this organization introduced the initial version of a formal project description document titled "A Guide to the Project Management Body of Knowledge," known in the industry as the PMBOK (pronounced Pim-Bock). This view of project management is generally considered to be the best-known and respected overview approach to this activity, although it does not describe how to do the processes indicated in the document. A similar effort in the UK created a comparable methodology known as PRINCE2, Projects in a Controlled Environment (Prince2). These two internationally recognized efforts and other similar ones represent a broad attempt to standardize project management and many of these include formal proprietary training and certification programs. Until recently, one could characterize the basis for these documents as modeling the project life cycle approach by assuming it to have a predictive scope and a cascading stream of defined work tasks; however, recent revisions are softening that highly predefined model structure view and that is a key activity to understand in the scope of this text.

Industry experience indicates that increased usage of such models has had a positive effect on project success, even though failure rates remain high in many areas. There is growing recognition that these documented project descriptions are bloated and cumbersome to execute, which results in extra-long delivery cycle times. In the author's opinion, there is also a feeling that the formal models are too static in their view and often require excessive documentation for no additional value, particularly in the planning phase. In examining the defined models one can conclude that project overruns are not caused by the model being wrong but more by the project not fitting the tenants of the model. In other words, if the scope cannot be clearly defined before execution, the rest of the model logic does not fit the problem very well.

Recognize that this statement now enters the realm of a religious type argument regarding how to best manage the project life cycle process. The conclusion from this historical review is the need to recognize the variability of projects and that none of the previous modeling efforts have satisfactorily embedded that concept into their design. The model needs to fit the problem and have flexibility with other processes. The thesis of this text is that the most popular management models do not fit this requirement but there are positive attributes that need to be extracted for use in another format. Nevertheless, this class of model represents the existing prevalent view regarding how to manage a project. More on this will unfold as we look deeper into the current views.

Quality Management Programs

After WWII, the U.S. began to witness operational success in creating improved products using quality management processes initially developed in Japan and copied in the U.S. to capture these ideas. Some example names for these programs are Zero Defects, TQM, ISO 9000, and most recently Six Sigma. These efforts focused organizations more on customers and efficient processes to deliver desired results. Important for this discussion is the quality movement's additional tool focus on organizational process improvement steps similar to a product-oriented project. The quality project management models are not as broad or complex as the traditional models but they highlighted and sensitized an increased focus on lean (no waste) and customer. Also, there is an implicit focus view regarding organizational learning and continuous improvement. As a result of this trend, the quality management school is now having both a tool and a cultural impact on how projects should be executed.

Strategic Project View

During much of the early evolutionary period, projects were viewed as discrete initiatives. As organizations began to formalize and proliferate this mechanism for "getting things done," it became obvious that organizational projects were competing with each other for scarce resources. To deal with this situation, many organizations created a formal oversight group to examine the collection of proposals and attempt to maximize their collective value. This activity is called Project Portfolio Management (PPM)—i.e., processes to evaluate the prospective value of competing proposals. A companion organizational decision process to this activity is often titled PMO or Project Management Office. These two linked functions combine to assess, prioritize, and approve projects out of the portfolio of options. The need for having formal processes to select project targets is now well recognized as part of the strategic management view and one that must be recognized in any modern formal model. Each of the key evolutionary items described here has had a significant impact on how projects are viewed and managed. This high-level project view represents an important thread of the management tapestry.

Universal Modeling Language (UML)

One of the more rigorous and interesting methodology attempts occurred in the early 1990s called UML. This attempt to design a technically correct approach to modeling software development was headed by three well-respected gurus in the software industry—James Rumbaugh, Grady Booch, and Ivar Jacobson. Their goal was to design a "single-unified method" (Visual Paradigm). By the late 1990s, this effort was felt to be the beginning of automated software coding based on sophisticated UML design models. Even though this model is still judged to be technically accurate, it is declining in use. There are attempts to link it to agile development but still no visible acceptance for doing this. This appears to be the case of a technically superior initiative sponsored by respected technicians that still was not accepted. The much less sophisticated agile development period started right after this time and was accepted by the industry. From a purely technical viewpoint, agile does little with rigorous modeling but focuses its attention on the project team and its lesser-defined sprint-oriented approach to producing outcomes. UML models are focused on software characteristics so there is little crossover seen in using these models for traditional product development, other than recognizing that schematic models can play a part in visualizing technical aspects. The history and evolution of UML are noted here mostly because of its characteristics regarding sponsors and technical sophistication, which logically are prerequisites for successful acceptance. It would be worthwhile for the reader to browse the comparative histories of both agile and UML, recognizing that both of these were focused on the same development target. Other than this history lesson, UML falls outside the scope of the integrated management model target.

Key Management Thought Leaders

Historically, the view of a project is related to the creation of some defined tangible product such as a building or other physical item. More recently, newer expanded views of a project have emerged as described in Chapter 3. Each iteration has brought new insights into improved management approaches and processes. There are five selected key examples of thought leaders that have changed the way projects are viewed:

- The quality school focus on customer satisfaction and lean processes, plus techniques to achieve those goals.
- The agile school of iteration and using sprints for team structures have had wide acceptance in various industry segments (more in Chapter 9).
- Elijah Goldratt's Theory of Constraints is a model based on a traditional design but with added processes to speed up output (more in Chapter 10) (Goldratt).
- The introduction of agile/Scrum methods in 2000 has had a significant impact on the management process by operationalizing the agile principles.
- A high-level umbrella conceptual model titled SAFe (Structured Agile Framework) was introduced in 2011 and defined a broad architectural view needed to support the increased use of agile methods for larger projects (CIO).

■ The portfolio view of project selection is now a recognized part of the management process and one that needs to be integrated into a combined life cycle organizational/project structure (more in Chapter 12).

There are also many other niche management-oriented models in play today, but the examples above represent a reasonable content coverage for describing new views toward the broad management conundrum needed to improve the project landscape. It is also important to suggest that most organizations adequately use only fragments of these processes. As an example, none of the models outline the role of a sponsoring organization in their support of the project, yet this has a significant impact on project performance. This external link to the project must be recognized as part of the management process.

The challenge we face in integrating new and improved management practices into the process is to prove that their value is worth the change in current practice. The first step in doing this will be to show where the current models are not optimum and how new processes need to be reviewed. If one accepts the statement that the current models do not match reality, it should be easier to look at modifications. We are now at the evolutionary stage of project management where one can recognize that projects look similar in their birth-to-death life cycle but have very different management needs based on their profiles. Given this recognition, the future strategy should be one of mapping the variable management needs to the unique project rather than forcing the project to fit into a fixed model structure.

Variable Work Structures

As indicated above, there is a tendency to pursue project management with a singular biased type view of the work to be performed. We will call this the current organizational belief culture. Five alternative disconnected views are useful in viewing project work. These are as follows:

Predictive—Cascading work groups and linked tasks that fit the waterfall design.

Iterative—Utilizing the agile school of thought with iterative sprint cycles of development searching for customer satisfaction.

Work chains—The view of work described by Goldratt's Theory of Constraints which views tasks as a coupled chain to be executed as quickly as possible. The analogy here is that the project should be managed without a fixed schedule and viewed as a track relay race with no schedule inhibitors.

Process-oriented—This class of project is focused on delivery systems with heavy customer interactions. These are often pursued using the iteration model or Six Sigma quality school tools.

Portfolio management—This evolving high-level project view involves the strategic aspect of the management process with a tight linkage to the ongoing set of projects.

Each of these work-type goal views contains valid management strategies that have merit to be included in a re-engineered and integrated management model based on a project's profile and goal characteristics. Recognize that each of these strategies should be implemented to accomplish the parameters defined in the project profile. Examples of these work selection strategies are summarized:

Predictive—best suited for tasks in which the work required is reasonably well defined. A predicted project plan can be produced with attendant control techniques. This is the best management architecture for forecasting and control purposes but suffers badly when the model assumptions are not met.

Iterative—work in this category is best suited for situations where there is active user involvement and a proficient team who collectively can evolve fuzzy requirements into the desired customer outcome. Experience shows that his model generates high customer satisfaction but suffers when tight control over schedule and budget is required. Here the work is broken into short "sprint" cycles where interim results can be verified. Some argue that this approach only deals with customers and is not sensitive to the management of schedule and cost variables.

Work chains—the Goldratt Critical Chain process for managing task execution is useful for situations where speed of completion is paramount. The overall model may be more complex than many organizations can handle, but there are subsets of this model that have great task management potential.

Process development—this type of work activity is characterized by examining human-based organizational processes and generally represents process reengineering efforts—i.e., improving customer service or process cycle times.

Six Sigma quality process. There is an emerging subset of standardized modeling tools suitable for this type of activity. In many ways, this fits the iterative model.

Portfolio management—the strategic management level requires the assessment of project proposals, including early estimates, business value, risk assessment, resource considerations, etc. This layer becomes the starting point for further project development activities that will lead to one of the domains outlined above for execution.

The process of homogenizing these views into a single management structure will require first that the reader understand what each of these basic models entails concerning work management. It is also important to understand how various management gaps in current practices that impact successful delivery can be improved with the usage of these components. In this case, it is not the model that is in error but the proper work management selection for the related process. In other words, given the character of a specific task work unit, what is the best way to execute it to produce its output goals—i.e., scope definition, delivery speed, control, resource management, customer satisfaction, or risk? More elaboration on this variable work selection idea will be offered in upcoming sections but this represents one of the core components of the integrated approach. An improved model view must have the capability to be flexible regarding the way various work units are executed based on the project profile priority goals. The expectation is that there will be multiple options within the project and not a single approach.

Even though the reader of this section may be knowledgeable in some subset of these model characteristics, it will be necessary to understand these aggregate broader views based on different work characteristics. It will be equally important to understand not only the design theory of the model but also understand the logic gaps in that theory based on the project characteristics. This will require an analytical approach to match the alternative models with their design flaws to the best work delivery strategy. We will also need to further describe alternative views regarding project success and how this new view may affect the management design of the project. The warning for such a discussion is that it will be a challenge because the reader will have to be ready to give up parts of their current practice and relearn how to structure their project according to these new ideas. There will also likely be disagreements on this collection of views.

Here is a thought question. What if the project schedule and budget are not the main determinants of success? For example, the competitive goal is to achieve a strategic market position, and the output product has to be widely accepted. This project needs a lot of future user involvement and evolving prototypes to review. Control is a lesser need than customer satisfaction. Because of the variable success view, it is no longer feasible to just pour the project tasks into a previously defined model structure and follow that to success. Recognizing this variable success criteria view will have a significant impact on the proper management approach. You are reminded again of the six blind men story in comparison to how this topic evolved as we move further through.

Delivery Forecast Projections

Accurately forecasting the future for any subject is difficult and yet that is the perceived goal of this text. As a data-oriented person, it is somewhat amusing to see weather forecasters sending out messages indicating that rain will start at 2:15 (six hours from now). Given the complexity of weather forecasting does this level of quantification make sense and is it accurate? Quite often, not only does it not rain that day but certainly not at the time predicted. Project managers often fall into this same quantification data minefield trap. In the project case, management requires that completion dates be defined, so the plan says that the project will be finished on June 4 and the budget will be $1,110,000. Neither of these occurs, therefore the project manager must not know what they are doing, just like the weather forecaster. Both know what they are doing but are reporting their data the wrong way based on the complexity of the topic. Given this, it would be better to get the receiving audience to understand the uncertainty and report accordingly. In the case of projects, this implies using probability distributions or range values rather than using discrete values. We'll leave the weather reporting issue for another group to solve.

A change in how projects are managed seems destined as other methods are now being documented as showing improved success, even though they also do not appear to contain a full breadth of view. In response to all we are describing here, project managers today are band-aiding their models in reaction to the user and senior management prodding for quicker completions or answers that don't make sense given the situation.

One of the common complaints about projects is the "excessive" time they take. Certainly, this variable is an important one and is a key success item. There are many examples where a project team took a shortcut in the interest of time which later resulted in project failure. All project managers would rather go slower, while all users would rather go faster. It is important to remember the *Tortoise and the Hare* story (remember the hare lost the race). Is it better to be fast and later crash the ship, or plan carefully and arrive safely a little later? The *need for speed* in the project world can be disastrous with a poor pilot guiding the process. There are many industry examples of projects taking quick silver bullets that did finish quicker but then failed because the new technology had some unknown flaws. There is no simple answer to this question, but it simply represents one more stressor that the project manager has to deal with.

An attempt to define a more flexible management model requires that current shortcomings are recognized and processes derived that better match reality. Our assessment of the current practice is that there is an ineffective understanding as to why the model is not working and an associated lack of understanding of what the proper project priorities are. A well-skilled project team may use a patched traditional model to achieve better success rates, but few project managers have this depth of understanding. Hopefully, the pieces outlined here will sensitize you to the various issues that can result from this. Much of what is described here will be not only where to look but how to effectively integrate the pieces into a viable solution package. We have begun to show various ideas related to some of the core success issues, but there is a lot more to understand regarding why projects fail and some of the management drivers that can help mitigate these factors. One of the currently accepted management specification models defines ten knowledge areas that are important (i.e., PMI's PMBOK). This is a useful background theory, but the question remains as to what degree these are important in a specific case. The question of how thorough we should be in any of these areas is a key success issue. It may be reasonable to skip formal risk assessment but is that wise? Many projects implicitly do just that and hope all will turn out all right. How does one decide on factors such as this? Which of the major management factors is most important?

These unresolvable tidbits are dropped here to make you think a little about why the answer cannot be a fixed cookbook or step. The future approach requires mature technical and management professionals who understand techniques to match different work profiles to the defined delivery goal. These priorities must be ranked for each project and not assume static variables. For example, concern for risk may be the number one priority, while the schedule may fall to number four. Admitting that all factors are not the same is just one of the cultural changes implicit in a new view.

The existing standard waterfall model design has significant advantages if a project meets the underlying assumptions, but even here there is a question as to how thorough one must be with the defined processes. Some might believe that the proper way to build a house is to collect a pile of lumber and other resources in the yard and work with the owner to decide what the house will look like, however, most would agree that some level of predefinition is best, especially if user satisfaction, schedule, and budget are also factors of interest. In essence, the philosophical question related to project management is based on one's belief regarding the planning rigor required in known target areas of interest (i.e., level of definition in scope planning, task estimating resource analysis, risk analysis, project approval process, status reporting requirements, stakeholder concerns, etc.). Higher levels of plan definition for each of these add additional time and cost to the project. As an example, the following decisions made regarding the rigor required from proper project management will significantly influence the appropriate approach:

- The level of scope predefinition before execution—the house-building example
- Deciding the level of work management delegation to the project team—this can impact scope definition and other management roles
- Specifying project deliverable priority goals and constraints (i.e., speed, cost, risk, customer satisfaction, etc.)
- Quantifying resource capacity needs (quantity and quality)
- Team management techniques to apply (morale, productivity, professional growth)
- Defined level of stakeholder support and communication

Projects now are viewed as any structured work activity that changes the existing state of a product or process. No longer does the output goal have to be a tangible product. Focus your mind on this broader view by envisioning a major reorganization project involving the movement of people, facilities, physical infrastructure, technical infrastructure, customer interfaces, and a host of other factors. This is taking a current state and moving it to a future state, thus a project. The new management model needs to fit this sort of activity as well.

This historical overview-related technique and philosophical section is the first step in focusing attention on the essence of global project management and offering clues as to where one might look for better methods to execute those ventures.

From this overview, you should see that the evolution of tools, processes, and documentation has been ad hoc at best. Many of the items developed were niche targets and not suitable for every situation. None of the examples would be considered to represent the universal best guidance regarding how to best approach a specific project. The key point has been made multiple times here that a fixed cookbook-type guidance model is not the answer to this complex environment. The future is going to require a more customized model based on the particulars of a specific project. Not explicitly dealt with at this point but highlighted now is the increased recognition of customer satisfaction as a project goal. That has been a somewhat secondary goal on the traditional delivery goal list in the past but it now is recognized as requiring much more active concern.

The wide variety of real-world projects and the various factors outlined here are intended to support the justification for proposing a new flexible model for the management process. This is going to have to include more than a listing of the important things. The future model must contain a broader view of the management process, including birth analysis, strategic and tactical planning, execution techniques, and customer links. The result needs to offer some prescription regarding how detailed each component needs to be pre-planned based on the project profile. The design approach must contain both a decision architecture and key success factor techniques.

A visible laboratory of project management practice lies all around us and these case scenarios are the best mental models to observe the impact of both good and bad techniques. Hopefully, examining this real-world view will make you more aware of poor management and from that create a more comfortable view of the process. That is the learning lesson that may well be the major reader benefit. For instance, does it bother you to wait on the phone for 30 minutes for *customer service*? What about sitting in a line of cars watching road construction that has traffic locked up for miles and seeing only a few workers in sight? Is it a good service goal to allow customers to wait for an hour to spend ten minutes with the doctor, or stand in queue line when there or other workers who could leave their non-time critical task to help? Once you get the spirit of project management being critical of bad designs it will bother you more. Also, you will instantly see what you would do to improve the situation.

It is time to start looking around at real-world projects and build a management model based on the ideas described here.

Conclusion

The early evolutionary period (1945–1960) legitimized the role of formal projects as the best organizational strategy to "get something done." Today, the term *project* is ubiquitous in our vocabulary. The management vision for this activity became defined as a sequential-step life cycle view, often grouped into macro stages related to defining, executing, and implementing the goal. This view remains the most utilized high-level definition today, although there are storm clouds on this horizon seeking new methods. Associated with the traditional view, projects are recognized to have a birth-to-death life cycle of decision and task-level execution work steps. Some require these to be clearly defined, while others are less formally defined. The thesis of this text is that project profiles need to be used to customize the management process using a broader set of flexible options to achieve the prioritized deliverables.

References

CIO. 2022. The Scaled Agile Framework (SAFE). https://www.cio.com/article/220569/what-is-safe-the-scaled-agile-framework-explained.html (Accessed July 21, 2022).

DCMA. October 2012. Earned Value Management System, DCMA-EAPAM200.1. https://www.dcma.mil/Portals/31/Documents/Policy/DCMA-PAM-200-1.pdf?ver=2016-12-28-125801-627 (Accessed May 15, 2019).

Goldratt, E. M. 1997. *Critical Chain*. Great Barrington: The North River Press.

Prince2. 2022. What Is Prince2? https://www.prince2.com/usa/what-is-prince2?gclid=Cj0KCQjw8uOWBhDXARIsAOxKJ2ENUip9QohrZbSI-x7fYkrflxRnEDSJ_qmgmYiiK9q-YH7QiaWM6akaAqLPEALw_wcBPRIINCE.Ref (Accessed July 21, 2022).

Visual Paradigm. 2022. The Unified Modeling Language. https://www.visual-paradigm.com/guide/uml-unified-modeling-language/what-is-uml/ (Accessed November 3, 2022).

Chapter 5

Project Success Drivers

Introduction

When one asks whether a project is successful or not, it is important to understand the audience asking the question. One group of stakeholders might say the project is very successful, while another might judge it as a complete failure. Both could be correct in their assessment and that is the dilemma that the project manager faces—i.e., which group to satisfy? On one side, the sponsor funding the effort is very concerned about the level of expenditure, so cost is their primary success measure. Second, when a project is approved it should have been selected to support an organizational goal. That can't be achieved until the deliverable is working, so time and functionality are the most critical variables to this audience. Let's look at a confusing fourth failure-type scenario to add further complexity to this question. An expensive nuclear power plant project is "successfully" completed and in operation. All goes well until a couple of years later when, in 2011, the Japanese Fukushima nuclear plant was hit by a giant tidal wave that washed over the installation and wiped out the plant. Is this still a successful project? The source of the tidal wave was an earthquake miles away that sent a wall of water to the plant. This unusual example takes us to the edge of the project success question. Another highly visible post-implementation example is the 1986 NASA Challenger explosion. This catastrophic event was reviewed by teams of experts to discern the cause of the explosion. In this case, the failure was traced to a simple O-ring that failed due to cold weather and management's decision to override the temperature specifications for this item. This operational result was based more on the management decision but the project still failed. These are just two well-known public examples of project failure after delivery. These two failure categories are rooted in different flavors of risk assessment and management decision-making. The former example failure root cause remains with the project domain even though one might argue that this was a very unusual cause and should not have been expected. The nuclear power plant issue could have been mitigated by simply moving the plant after recognizing the ocean threat. The challenger case is more complex than that and simply suggests that the issue of risk is one that needs careful attention even after the project is completed. Collectively, this group of success/failure examples highlights the important role that risk assessment and management have in the overall project life.

As indicated above, success views and quantification vary by the audience. From that observation comes the question of what variables can one use to measure success. In the power plant example, a better risk assessment might have uncovered the tidal wave potential and from that a decision to move the location and save the operational disaster.

The classic view of project success is to compare planned values of schedule, cost, and functionality versus actual results. But even here the question is how much variance in these factors is still considered a success (10%, 30%?). Variability in survey mechanics makes cross-comparisons suspect. For this reason, one should take each of these sources with a "grain of salt" as the old saying goes. Accurate data from non-public projects are very hard to collect and even public projects are difficult to audit. Measurement rules vary, and the person evaluating the results may have a unique bias. Nevertheless, we have a starting place to discuss the significance of this topic.

Industry Studies

There are numerous published studies related to project success, but four specific sources are reviewed here and are worthy of further detailed reading. These sources are the Standish Group (stnadishgroup.com), Project Management Pulse Surveys (PMI 2016) and (PMI 2017), Stanton's 2011 dissertation survey (Stanton), and Nelson's timeline study (Nelson 2007). Each of these sources offers a critical summary of the project success story, but none of these offers a clear prescription regarding how to achieve success. However, they do provide insight into what the major failure factors are. These describe the WHATs but much less on the HOWs. The one theme that comes through these studies is that projects are not overly successful regardless of how one measures them. Beyond this, these studies have highlighted the complexity of project execution and the breadth of factors that cause them to be less successful. The sections below will summarize the research findings but the reader should review other focused industry data to understand how the causal factors change.

Standish Surveys

The Standish Group was one of the early IT survey organizations that quantified the extent of project failure rates in the IT industry (Standish Group 2020). Their annual CHAOS editions of these surveys provided new insight into the magnitude of this issue and also began to add priority data on the various root cause factors.

Figure 5.1 shows traditional success and failure measures for IT project performance from 2011 through 2015.

Industry recognition of this data brought increased awareness to a wide audience and clear recognition that projects had poor success results. There are at least four key revealing performance metrics in this survey:

1. Large projects have very low success metrics but also indicate less failure grades than other groups
2. Small projects had the highest success rate but still did not have exceptional performance
3. Average failure rates were in the range of 11% to 31%
4. The three most prevalent factors correlated to success/failure were senior management, user support, and clear requirements.

	SUCCESSFUL	CHALLENGED	FAILED
Grand	2%	7%	17%
Large	6%	17%	24%
Medium	9%	27%	31%
Moderate	21%	33%	17%
Small	62%	16%	11%
Total	100%	100%	100%

Figure 5.1 IT project performance (2011–2015) (grouped by size). Source: Standish Group with permission

These findings would be considered the historical base case for project status.

More recent editions of the CHAOS survey (2020) began to show the use of alternative development approaches, notably from the use of agile/SCRUM methods. Reports from this dual development environment are not consistent with past traditional delivery models, so one has to be careful in interpreting data across management model types. (Standish Group)

It is interesting to note that this later survey made the comments that there was questionable value in the role of a project manager since the SCRUM approach tended to be more of a flat team approach. This illustrates another issue of examining project success. In the case mentioned, another development method was used and this recommendation does not fit all project types. More details on the CHAOS annual surveys can be obtained through the Standish Group website with a membership fee.

PMI Pulse Survey

As industry surveys evolved, they began to include more details regarding success or failure factors. As a later example, the 2017 PMI Pulse survey highlighted the following (PMI Pulse Survey, 2017):

■ Projects hit the major business target in the 60–70% range—this does not mean they succeeded but were focused on the right target
■ Projects on average fail in the teens (approx. 18%)
■ Projects finished on schedule and budget in the lower 50% range

There are at least two consistent results found here. These are:

1. Small projects succeed more than larger ones
2. The level of project failure by any category shown is still at least in the 18% range and some evidence that other environments may be higher.

One of the most troubling conclusions from the survey is the statement that approximately 12% of resources invested in projects are wasted because of poor performance. The study also indicates that less than 40% of the organizations placed a priority on creating a project culture that led to improved performance. The basic conclusion of this survey was that good project management improves success and lowers risk, which in turn results in delivering better economic value to the organization. A final point stated that mature project environments delivered significantly better results than less mature ones. There is more comment on organizational maturity later in the text. One noted value of the PMI data is the breadth of the survey audience and project types. These data results were gleaned from 2400 practitioners. Some of the survey details described the various management approaches used and these also add insights into the strategies being pursued.

Nelson Survey

One of the broadest and most extensive survey reports is credited to Dr. Ryan Nelson (Nelson, 2007). This survey looked at the question of success/failure through factors more than individual results quantification. The following three key project success factors were identified as most significant:

- Executive support
- Stakeholder management
- Risk management

Once again, the top two items are consistent across the Standish and PMI Pulse surveys. Even though the risk was not so prevalent factor in previous surveys, the examples given earlier for the nuclear power plant and Challenger explosion certainly add credence to the need for better examination of the internal risk factors.

A lower level of detail in the Nelson survey quantified 36 more detailed factors. While this list is interesting, it is overwhelming in size. Out of the defined issues, 14 of these were present in at least 20% of the cases (Nelson 2007). The top five factors according to frequency are:

1. Poor estimating and scheduling
2. Ineffective stakeholder management
3. Insufficient risk management
4. Insufficient planning
5. Shortchanged quality assurance

Four out of the five items above are labeled as management process shortcomings. The following is a brief scenario view of these factors and how they affect outcomes:

- *Technology decisions*—there is a tendency to find use a new tool or untried technology that will improve output. These are called silver bullets and they often fail to achieve their promise.
- *Resource management*—not having timely resources or adding resources mid-project to improve the schedule can both negatively affect results.
- *Scope creep*—failure to properly manage scope changes can be a chaotic management situation.
- *Inefficient planning*—this factor is related to coordination, resources, work management, and related processes

Approximately 85% of the identified problem factors were equally balanced between processes and people. Also, the top three factors occurred in approximately one-half of the surveyed projects.

One of the data collection processes used in the Nelson survey uncovered yet another surprising success perspective (Nelson, 2007). These data revealed that projects initially judged successful at project completion using the traditional three-factor measures did not correlate well with user views three years later. There are multiple reasons why this might occur. First, the users may not have understood the requirement well enough to initially specify the functionality. Second, given the traditional push to complete a project on the plan measures, the level of change requests was either cut off or minimized. Regardless of the root source of this unusual finding, it seems clear that closer user involvement in the requirements definition process is a critical success element.

Stanton Survey

Michael Stanton published an extensive doctoral dissertation on the general topic of project success factors. Although the title of this work was focused on the role of project selection, it included all aspects of the delivery cycle. This data had a broader view of success with these three additional views (Stanton, 2011):

- Was the project done right?
- Was the right project done?
- Were the right processes done repeatedly?

In some ways, this view of success says that you should select the correct target, pursue the target using best practices, and do this every time

Concluding Thoughts

PMI's 2017 Pulse-of-the-Profession survey concluded that the following four characteristics led to success (PMI, 2017):

1. Clear and doable project goals
2. Careful planning
3. Adequate resources
4. Stakeholder management

One prescriptive conclusion from this research was that the existence of a formal Business Plan and project plan were the two key artifacts for success. From reviewing these various industry approaches that claim to quantify project success and failure rate, we should now be convinced that there are many ways of looking at this topic and there are numerous factors affecting that outcome.

There is no easy way out of this definitional quagmire. The obvious goal is to just satisfy all of these measures, but that is impossible. The one management requirement that surfaces is the recognition that this topic must be included in the project design architecture and specification.

If a project is overrunning its plan and not meeting interim milestones, it may be best to cancel the effort, or at least redefine the goals. In order to deal with this dynamic situation, all key stakeholders need to be involved with both plan and actual status measures. If one looks at the three basic performance measures, it is potentially possible to adjust some process variables to achieve a higher-priority deliverable. In other words, cut scope, increase the budget, or even extend the schedule in order to better match the defined success priorities. The tracking process to evaluate characteristics of project success needs to be recognized as a dynamic decision and not one that can wait until the project is over. It is important to understand the constraints over which a project is judged a failure and the flexibility or variance that can be tolerated and still be considered a success.

Defining Project Success

The increased visibility of surveys quantifying project performance is viewed positively in one way. That is, it has sensitized the industry that things were not going well, but on the negative side can now be viewed as potentially misleading. One aspect of this is the excessively long list of factors to deal with. As with any quantification process, questions have now emerged as to what constitutes success or failure and how that item was defined in the survey. The answer to that is variable and depends on the particular source. Understanding this management component has been made worse by the expanding view of a binary rating. It should now be defined as a vector of many possible elements.

One frequent discussion point that goes with measuring a project's success involves whether any particular approach to managing the projects had anything to do with positive or negative results. From the intuitive side of this, one might conclude that the maturity of the project team might well be a major factor in the resulting outcome. But the various survey data seems reasonably consistent to the degree that another conclusion is that the results are independent of methodology. The most troubling metric is the consistent reporting that failure rates have not declined over the past ten years. At the very least, this suggests that improvements to the management process needs to be made.

Improving the Outcome

Associated with quantifying the results of projects there have been multiple attempts to describe cause and effect factors. The one very clear correlation between success and failure comes from project size. Small projects succeed at a significantly higher rate than large ones. Not every project can be made small, but packaging work into smaller groups or phases is often possible. There are four reasons why small-size projects are more successful:

1. Requirements easier to define
2. Quicker to produce (less time for changes)
3. Smaller team size (better communications)
4. A smaller number of stakeholders involved

Notably, there is a strong consistency in the performance factors identified over time and this suggests that organizations are not working adequately to improve their results. The causal factors have now been weighted as to frequency, and based on this weighting scheme, approximately 70% of the factors affecting success (or failure) are related to the following five factors:

1. Senior management support
2. Organizational support of the project
3. User support
4. Clear objectives (requirements)
5. Skilled resources

Note that this list does not contain project management skills nor a particular development methodology, although each of these factors is listed lower in the ranking. One way of looking at why these five items are static success limiters is that they are not clearly understood as to their significance. Regardless, the stability of these on the list leads to a strong recommendation that the management process rule should be to "watch out for these things." The fact that the same factors repeat over time may mean that "doing it" is much harder than it appears. There is one other global operational variable buried inside all the visible factors that likely dominate the overall success factor equation. That is *communication* across all elements. Finding the right operational methods to support good communication across the various project groups may well be one of the key answers we are looking for.

Evolution of Agile

The 2015 survey report from Standish provided comparative success result data for agile (iterative) versus the traditional waterfall (predictive type) methods. Figure 5.2 contains the results from the Standish survey. Similar less auditable results data of this type has energized a development methodology discussion to question whether a particular management process can improve results. If one were to accept this data at face value, it would suggest that the industry has been following the wrong model all these years. Note that agile outperformed the waterfall approach in every category.

SIZE	METHOD	SUCCESSFUL	CHALLENGED	FAILED
All Size Projects	Agile	39%	52%	9%
	Waterfall	11%	60%	29%
Large Size Projects	Agile	18%	59%	23%
	Waterfall	3%	55%	42%
Medium Size Projects	Agile	27%	62%	11%
	Waterfall	7%	68%	25%
Small Size Projects	Agile	58%	38%	4%
	Waterfall	44%	45%	11%

Figure 5.2   Agile vs waterfall results. Source: 2015 CHAOS, The Standish Group

Data of this type glosses over a very complex set of variables but the industry reaction shows the impact of surveys n perceptions. Agile is the most commonly mentioned target for this new development approach but even this designator is questionable given the various dialects of the methodology. (More details on this are covered in Chapter 9.) Also, note that the comparative data were primarily collected from software projects that supported an iterative prototyping approach that does not translate well into more product-oriented project types. Still, there is sufficient data to require more research into why this is perceived to be a superior method.

Conclusion

Industry surveys have enlightened the industry into what appears to be a marginal management environment, Failure rates are accepted as being too high and there remains a question as to why the same five factors continue to be recognized as sources for failure. For purposes of future text sections, we will take the average value of 50% as the industry project failure rate. Also, the surveys have provided a wealth of factors believed to be related to this negative performance. The following list makes a good summary of issues to deal with as we move forward:

1. Communication is one of the major management gaps.
2. Getting accurate requirements is a common root cause of poor outcomes.

3. There is a diverse dichotomy of success objectives across senior management, the project team, and future users, which complicates the success evaluation.
4. Having a productive team is a key success factor.
5. The value of "lean" over fixed bureaucracy is emerging as a proper change direction.
6. A mature host organizational support environment can improve project results.

There is a continuing reinforcement that the management process is difficult even when the factors leading to failure are defined and no generally accepted delivery process guarantees a successful outcome. On the other hand, failure to focus on this topic can lead to very undesirable outputs, and for that reason dealing with the known success limiters should be part of the best practices regardless of the project type.

References

Nelson, R. Ryan. 2008. IT Project Management: Infamous Failures, Classic Mistakes, and Best Practices. *MIS Quarterly Executive*: Vol. 6: Iss. 2, Article 4. Available at: https://aisel.aisnet.org/misqe/vol6/iss2/4

PMI. 2016. Pulse of the Profession 2016. http://www.pmi.org/learning/thoughtleadership/pulse/pulseoftheprofession2016 (Accessed November 15, 2022).

PMI. 2017. Pulse of the Profession 2017. https://www.pmi.org/learning/thought-leadership/pulse/pulse-of-the-profession-2017, (Accessed November 15, 2022).

Standish Group. 2020. Project Managers Fail to Help Software Projects. The Standish Group. https://vitalitychicago.com/blog/project-managers-fail-to-help-software-projects-standish-group-chaos-2020/ (Accessed August 5, 2022).

Stanton, Michael J. 2011. Portfolio Management: Perceptions of the Project Manager, Ph.D. dissertation, Capella University.

Chapter 6

Project Externals

Introduction

There are three visible areas related to project variation—i.e., risk events, scope changes, and variations in work estimates.

Figure 6.1 schematically shows how these relate to the defined project domain.

This chapter will focus primarily on the two-project external events related to risk events and scope changes. Many project managers would say that improved handling of these two external Workplan items had a significant impact on variability in the deliverable results. Recognize that neither a future risk event nor scope change data should be included in the base plan. However, related data does need to move into the product delivery plan once they occur and then be recognized as part of the work requirement. Common practice often does not handle these items properly and failure to do so can corrupt the integrity of the base plan. The following describes the basic mechanics for proper handling of the risk and scope change events:

- Future risk events should be funded through an external contingency fund. When one of these triggers, this fund will be used to deal with the event and those funds will be transferred to the base plan.
- Approved scope change will be funded through an external scope management reserve. When a scope change is approved the estimated cost of inclusion is extracted from the reserve fund and transferred to the base plan.
- A project schedule buffer is attached to the original base plan to handle various increased times created by these additional work requirements. Recognize that this is not an ideal solution. A better arithmetic answer would be to add these schedule increments with each risk or scope event but this may be administratively infeasible. The key point is to recognize their occurrence outside of the defined work segment.

DOI: 10.1201/9781003431091-7

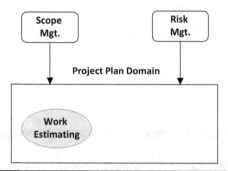

Figure 6.1 Project externals

The risk and scope management processes are frequently mishandled items and embedded in a padded view of the defined plan. This practice hides these events and thereby loses visibility to their impact. Contrarily, the approach recommended here is to handle both of these external project dynamics through the use of a separate external fund for scope change and a contingency reserve to cover future risk events. Both of these would be packaged with the approved base plan but managed separately for control purposes. Think of the management logic this way—if there were zero scope change and zero risk, the base plan would not need alteration and a management reserve would not be needed. Alternatively, when there is a status change between these two areas, there is no additional place to fund such activity. Both of these events may potentially occur but the exact degree can only be probabilistically defined. The essential management idea related to these is they are not in the defined work plan until they occur. When the event triggers, the reserve fund is tapped to pay for the event, and related resources are moved into the active project view. This is an essential concept for the proper management of these two items.

This area of project management is not simple to understand or agree on a solution. For example, if you don't know how much scope change will occur, how do you size the reserve? The same question exists or risk. Unfortunately, we can't exactly quantify the reserve values for these two questions, but regardless of the reserve size used, at least, we can track the magnitude of the two items and maybe do a better job on the next cycle. Failure to isolate the reserve from the base plan camouflages both factors and that hinders the ability to manage the process.

Changing Work Requirements

It is common for new insights to occur during execution, and in many cases, these will be added to the initial project view as new work requirements. Future chapters will add a more detailed description of the actual handling process mechanics for these two items. The important introductory point regarding requirements changes is that they represent newly approved work and these events do not technically represent a variable in the project plan until they are known. Yes, such changes will often increase both the budget and schedule above a perfectly estimated plan, but when they do occur, they represent newly defined work that must be shown in the active plan. The key question here is how to estimate the magnitude of these ill defined events. If the change was approved by management, it seems logical to say that the additional cost and schedule should be added to a modified plan. The project team should not be blamed for an apparent variance created by this action. From this brief process description, let's examine a basic philosophy for handling these events.

The first management requirement for a project is to focus on the defined requirements—i.e., planned deliverables, schedule, budget, and resources. Newly approved work will be evaluated externally and if approved will be added to this initially approved plan, which should now be operationally recognized as a modified project. Too often, planned versus actual comparisons do so using the original plan and erroneously treat any variances as indicators of project overrun. Growth in a project because of newly approved requirements is not an overrun if management has approved such changes. This is a subtle point but failure to keep this activity isolated and managed also creates other bad practices. To isolate this category of work, a scope reserve to fund these changes needs to be attached to the base plan. All approved scope changes will be funded out of this reserve. A second even more subtle point is that the change control analysis process should also be funded out of this reserve as this represents supplemental work. We have never seen that done. Essentially, the goal here should be to isolate all actions related to changing requirements. For projects using third-party contract vendors, this process is often handled by formally approved resource supplements to the base contract, but this formality is often missing in internally staffed initiatives. In any case, recognize that all of this class of activity is extraneous to the original requirements and should be treated as such. If changes are not allowed, this process would not be needed, but this is a required activity in most cases. The risk management path will occur on its own in similar fashion. The logic for using an external budget reserve is to provide better control of the base plan and to provide insights into the magnitude of this external source of overruns.

One remaining management question involves how to establish the size of a *scope reserve.* This reserve should include both the estimated budget and schedule amounts associated with scope change activity. As these fund resources are allocated the approved plan level is adjusted upward accordingly. As an example, if a change is estimated to require $10,000 and add three weeks to the project plan, this is accommodated by extracting that level of funds from the contingency reserve and the new work referenced in the new approved active work plan. Budgetary amounts related to this class of activity are easier to define than related schedule increments. In many cases, an approved change is too small to estimate task schedule impacts. A typical approach to this is to attach a single schedule scope change buffer to the project schedule. This buffer would reflect both an amount for task overrun and scope change impact. This solution is not completely clean but at least it shows the budget impact of scope change. The key logic of this process segregation approach is to isolate the change dynamics from predefined work and provide a better ability to evaluate the status of the work plan. This approach to scope management will require some additional discipline but is a worthwhile process.

The question now is what causes requirements to change? As an example, a subsystem test could uncover an issue that requires extra work to correct. Also, a missing work task or uncovered design error could be the cause. Beyond this, there is a myriad of reasons why a better understanding of the requirements emerges. The operational challenge in evaluating a change request is to decide if the additional cost or time is justified. Sometimes the existing requirement is good enough but the stakeholder is now looking for something nice to have. (Recall the acronym MoSCoW described earlier for scope definition parameters.) One can look at this aspect of the overall delivery process as a tradeoff of budget and schedule versus improved functionality. These often become management questions to resolve. On the negative side, changes made to the work plan during execution can create chaos in the delivery management process so it should be managed accordingly. In some cases, a change can require not only additional resources but can also negatively impact other sections of the project or work previously performed. Both stakeholder and technical units must be involved in these decisions. In addition to the control aspects outlined here, there is a negative organizational cultural issue related to it. For projects that are executed with primarily internal resources, there is a tendency to just pad time estimates to cover undefined future changes. The problem with this approach is that it hides the magnitude and root cause of the change as work unrelated to actual task estimates does not relate to real events. It is important to isolate this activity and recognize the impact by separating it into an isolated management domain for analysis and control. In the traditional project world, this is called *Scope Change Control* but the idea should be considered for any project that introduces requirements changes within the delivery life cycle.

Dynamic scope changes are a threat to destabilizing the project plan, so this management area must have proper focus from both the technical and stakeholder side. A formal decision process is needed and no changes to the defined requirements should be made without approval from the formal process. Think of this process as a mini-control project in its own right.

Risk Management

A second troublesome external factor affecting project performance comes from risk events that are difficult to predict or plan for. Dealing with project risk may well be the most misunderstood and complex management issue in the life cycle spectrum. Let's describe why this statement is true. First, a risk event is defined as a probabilistic *known/unknown*, meaning that it may happen, but its actual occurrence probability, timing, and impact are very difficult to quantify.

A complex yet interesting real-product case study regarding risk offers insight into this management phenomenon. During 2018 and 2019, Boeing, historically a very mature risk-oriented company, and the entire aircraft industry were negatively impacted by the introduction of a new model of the 737, MAX 800. On the surface, this model was simply an extension of multiple successful similar upgrades over twenty years. Suddenly, in 2018 and 2019, two crashes occurred on international flights resulting in the death of 346 people, and from this, all of these models were grounded for several months (Perell 2020). Subsequent analyses over the next several months uncovered the root cause of these crashes to flaws in a software-automated flight control system. If a pilot did not manually correct this unrecognized automation flaw, the plane remaining on autopilot could stall and crash. If one looks at this issue mechanically, the problem was a design flaw that should have been recognized through initial system tests and evaluation processes. The design error could also have been caught in early flight testing and a relatively simple fix made before government entities got into the picture. This example represents the potential magnitude of unfound risk events that negatively impact the value of the project. In this case, the event is described as the tip of a corporate risk culture gone wrong over time. Perell offers a much broader view that claims the organization began to focus more on profits and the expense of risk management (recall the rubber box example from a previous chapter). Space constraints here do not allow for further elaboration of the organizational background culture change, but this example does show what can happen when the project risk environment is not properly dealt with. On the surface, it appears that this event could have been identified and resolved during the test cycle with minimal economic loss. Also, note that the product scope definition was accurate and the project was successfully executed in a normal fashion with high customer acceptance.

In 2003, the 737-model family represented 25% of all large commercial planes. Failure of the MAX 800 resulted in billions of dollars of damage to Boeing and the airline industry. The reader should seek the referenced Perell article to see more examples of the risk culture in an organization. In this example, the identified risk event was just the tip of a large cultural iceberg whose roots can be traced back to a declining organizational risk focus culture that eventually impacted operational delivery goals, technical skills, and even the organization's reputation. This example also shows that even major risk events can be hard to identify even with familiar products such as this. In this example, the risk event should potentially have been identified either in the design phase, basic engineering level, or the testing phase. There is a universal feeling that risk events should have been obvious after they have occurred. Also, this was a large and high-technology project so a more rigorous risk analysis should have been undertaken before the design approval and during final operational testing. Actions that uncover major risk events before execution help to minimize the potential impact of such an event if it were to occur later during the execution phase, or at the operational level as described by the MAX 800 scenario. Each step in the life cycle exponentially increases the impact of an unidentified risk event.

Recognize that a risk event is not preordained to occur. Even if they might occur, there are decision options that can help minimize or eliminate their impact. Here is an illustrative example of a risk event that can be changed by defined mitigation decisions. Assume that your office building does not seem to have adequate fire protection, what should you do? Nothing? You could decide to move to another better-constructed building and almost eliminate the current fire risk. Alternatively, you could install sprinklers, or decrease the risk level by establishing an active fire culture with internal resources trained to handle such events, etc. These actions might not eliminate the root cause of the fire but should lower the impact. Also, there might not even be a fire so even this level of concern did not add value to the outcome. The question here is how much time should be spent examining things that may not occur, which is taking resources away from making progress on actual deliverables. Based on this view, one can see why there is pressure to minimize risk analysis. This is the management conundrum surrounding how much effort to spend on risk and that is the essence of the management problem regarding the risk area.

The second view of risk management comes within the life cycle. In this segment, risk events could emerge from a design failure, loss of a key employee, or any other unplanned events affecting the project. The third area where risk impact can occur is as described with the MAX 800 project as the product is operated in production. At this point, all previous management processes have failed. The customer has accepted the results as a successful project and it is moved into production mode. The tradeoff in deciding when to not spend resources on formal risk management activities versus facing a catastrophic potential future impact of some failure event keeps knowledgeable project managers awake at night. One final philosophical point. No process can forecast the risk factors or level of exposure, but one important requirement is to not ignore this topic when you are under stress to move on. In addition, a post-project analysis is an important management activity too often ignored. Improvement in all aspects of this activity can be made by looking at the final positive and negative results.

To summarize the above, risk events can be most efficiently dealt with if identified before execution, but this is a time-consuming activity. Risk events occurring during the execution cycle often cause band-aid fixes that are difficult to implement. Finally, risk events that occur in production can be catastrophic as the examples have shown. The risk process operational approach described here is to use the same reserve concept as described for scope change. One of the early management questions to decide is how rigorous the initial risk assessment process should be. Even in situations where there is a long experience with the product or related technology, there can still be a new risk waiting. The Boeing case shows that one cannot be complacent even in a follow-on project like the MAX 800.

Developing a Risk Culture

Humans ignore risky events by assuming they will not happen to them. As an example, most humans learn to drive a car, but don't know how to change a flat tire, which is a highly likely event in your driving career. Also, we drive cars on congested highways with high accident rates but learn to ignore the associated risk. Some do this by driving the speed limit and obeying traffic rules, while others ignore all parts of the road risk equation. Probably the best way to evaluate your risk management culture is to examine your level of preparation related to the car you drive. Recognize that your involvement with a car e will trigger a risk event sometime in your driving career. At this future time, you will be likely inconvenienced when the event does occur. The first test is to ask is have you considered this as an explicit risk event. If you have done this, what risk event did you prepare for? A flat tire is the most logical. Do you know how to deal with this away from help? If you need a wrecker, do you know how to contact one at least in your home area? (Yes, Google can probably find one but is cell coverage also a risk?) Experience suggests that most people do not think about such risk events until they happen. Project managers cannot have that casual mindset and be successful. A significant characteristic of project success is to be prepared for unknowns. Risk management models are still immature, but having a risk-sensitive culture is mandatory and will provide needed support.

Both risk and scope change create new work requirements for the project that are hard to plan until they are defined. The following steps summarize the recommended approach to handling risk events during execution:

■ Probabilistic risk events are not shown in the approved base project work plan since they do not yet exist; they are managed through a separate risk reserve attached to the plan.
■ Establishing the size of the risk reserve follows the same logic as described for the scope reserve. Historical experience and rough estimating are the typical approaches.
■ When a scope change or risk event occurs, resources are extracted from their respective reserves and moved to the project plan to fund handling the event.
■ A post-project analysis of these items will help to improve future actions.

The Project Management Institute's PMBOK describes a formal risk management model (see Richardson and Jackson, 2019, chapter 22). There are numerous other sources for the reader to obtain more details and insights into this set of formal planning analysis mechanics, but it is a reminder that some sizing assessment is needed for each specific project based on its defining characteristics. In many cases, a less formal approach is adequate but some defined assessment level should always be considered. Regardless of the process chosen to evaluate the project risk, some undefined risk events will eventually occur and the expected view is that they will surface at the worst possible time. This well-known view is labeled Murphy's Law.

Experience with formal risk identification and evaluation models indicates that they are of marginal value in defining the actual events or impacts that will eventually occur. Research has indicated that actions related to risk identification are the least mature of all life cycle management processes. In many situations, risk events appear to be impossible to predict accurately. Does this mean that the whole area of risk should be ignored? Here are the risk views that need to be understood for any project:

1. A risk event that occurs within the life cycle or later can destroy the value of the project.
2. The maximum level of risk tolerance is a topic area that must be considered and formally discussed with project sponsors.
3. The mechanics of handling risk activity needs to be kept separate from the base plan management.
4. Tasks should not be arbitrarily padded in anticipation of potential risk.
5. The concept of defined risk reserves is an important aspect of managing this project area.

Risk Planning and Control

It is generally not possible to control something that has not occurred but that does not mean the appropriate way to deal with it is to just ignore it until it happens. There should always be a watchful eye approach and sensitivity to the possibility that something can go wrong. Formal control is focused on the approved plan and tasks are not adjusted to cover these events until they have occurred, then they become part of the active work plan. Many formally defined potential risks will not occur and the ones that do occur may be quite different from what was anticipated. Some risk events can be at least categorized and looked at as a group event, while others fall into the category of "unknown/unknowns" meaning they were not even considered as likely to occur. The previous power plant tidal wave event might be an example of this. Regardless of the type, the recommended management strategy for risk will be to use an appropriate level of activity to evaluate the risk environment and define how these should be handled either before or after they occur.

Even more important than the mechanical aspects of risk management, an organization needs to develop a culture of risk. Too often this topic is pushed aside in favor of cost-cutting, profits, and speeding up the life cycle, only to find out later that this was a poor long-term choice. Formal risk owners need to be assigned to various task areas to improve quick responses to these events. Their job is to be the front line of defense to quickly recognize the occurrence of a triggered risk event and take appropriate action. Slow recognition and reaction can exacerbate the impact of these events. Implementing an active organizational culture related to risk is often more effective than using formal assessment models or checklists, but the specific project profile should dictate how this topic is operationally dealt with. Our goal here is not to go further into the mechanics of this topic area but to highlight the negative potential and general handling concepts. This can be a very troublesome aspect of the management problem.

Learning Organization

Throughout this text, there will be a recurring theme of a learning organization—every day the organization needs to evaluate events and work status to adjust methods to improve future performance. That includes both the organizational culture and work process view. The topics of risk and scope management make a perfect case study for this. As these items emerge during the life cycle, they can be very disruptive and inefficient to absorb; however, hiding this class of events by padding tasks within the action plan also hides away the ability to assess how these events affect the project. If the role of these elements cannot be identified for analysis, it will be difficult to improve future performance. Keeping the two areas in the *sunshine* will help to learn more about how best to manage them in the future. If this area is not handled as described, a post-audit cannot evaluate the actual work status of tasks and the magnitude of any risks or scope changes that impact that performance. Through time and experience, these two complex areas can be better estimated and managed. This is the essence of a learning organization's goal. Successful handling of risk and scope is an essential component of project success.

References

Perell, David. 2020. Why Did the Boeing 737 MAX Crash? Long Form. https://perell.com/essay/boeing-737-max/ (Accessed June 12, 2022).

Richardson, Gary and Brad Jackson. 2019. *Project Management Theory and Practice*. 3rd ed. Boca Raton, FL: CRC Press.

DELIVERY STRATEGIES

This section of the text covers a series of classic delivery models followed by the introduction of a flexible integrated management model that encompasses targeted delivery techniques that have proven value in classic models. Following a review of three classic models, multiple success-oriented work management strategies are outlined for use in the integrated model. Also, a collection of success-related component processes is outlined. This collective technical background represents the elements that need to be combined into a working management process. The latter chapters in this section will define the model architecture and various other aspects related to its implementation. The resulting model can be mapped onto any project type and deliverable goal.

A brief tutorial of new work processes follows the model description. The last chapter in this section discusses some background of logical steps that lead to the new model. This provides a reasonable introduction to how the model might be accepted and what organizational culture issues exist.

The following list contains a brief summary of the chapters in this section:

Chapter 7 Project Delivery Models and Processes—This chapter introduces the management views which have influenced the design of current delivery models. The two main threads that will be followed in the text are the predictive and iterative views.

Chapter 8 The Classic Predictive Model—This delivery model is focused on a defined scope environment. The classic waterfall model is described in this chapter, including design assumptions, maturity usage, and delivery support tools.

Chapter 9 The Iterative Development Model—The iterative models described here are classic agile and the Scrum dialect. The design assumptions are described as well as delivery gaps.

DOI: 10.1201/9781003431091-8

Chapter 10 The Critical Chain Model—Selected attributes of this model are described to show how these design assumptions can compress the project cycle time.

Chapter 11 Organizational Support Architecture—For a project to be efficient, it needs support from the host organization. This aspect of project support is one of two defined components in the integrated model.

Chapter 12 Portfolio Management—The process of selecting the correct project for execution is described and roles outlined for both senior management and the related technical planning process.

Chapter 13 Integrated Model Design Components—Each of the previous chapters identified some key aspects of producing successful projects. The chapter collects those and begins to formulate a skeleton structure on which the new work execution techniques can be installed.

Chapter 14 The Integrated Delivery Model—This chapter describes the architecture of the new integrated that encompasses the best practice methods described in previous chapters.

Chapter 15 Modified Management Processes—This chapter outlines key processes that are modified to fit the integrated model architecture. Key changes described are flexible work queues, modified task estimating, and new status tracking processes.

Chapter 16 Integrated Model Tutorial—This chapter describes an expanded description of various processes embedded in the new model and provides details on various success drivers.

Chapter 17 Model Background and Installation—This chapter is a post-log to the new model in that it outlines the evolving thought process through time that led to the model structure. A second segment *describes* some of the installation issues that an organization would face in implementing the new work processes defined in the model. The impact on traditional status tracking is a major focus topic.

Chapter 18 Success Recipes—This chapter contains a selected listing of processes that are common factors in improving project outcomes. This is formatted as "Success Recipes" in a prescriptive format.

Chapter 7

Delivery Models
and Processes

Introduction

This chapter introduces the process of exploring how various project management gaps need to be better dealt with. Also, an overview of multiple work delivery characteristics provides an approach to specific work unit selection options. These options are an important part of the hybrid solution that allows multiple options within the same plan structure. An introductory view outlining the global components of an integrated model begins to surface here. The concluding section describes the next steps in the model description.

Delivery Options

An overview of current project delivery strategies could fill many books, but the goal here is to evaluate the main threads of delivery models that have received positive reviews. Each of the models falls into what could be called a key design assumption that drives how it then views the project work requirement. Also, each of these classic models implicitly claims to solve some perceived management issues. To their credit, all three of the classic models reviewed in the upcoming chapters are worthy of understanding for broader use. These model reviews will show their underlying structure and potential merits for further use (here we are replicating the blind men and the elephant story again). The list below summarizes the assumed environment for each of the main design categories. Four project management design assumption groups summarize the primary model designs:

DOI: 10.1201/9781003431091-9

1. Predictive—project requirements can be redefined and related tasks can be reasonably estimated. A project goal is often a tangible product.
2. Iterative—Project requirements are not well defined. The basic assumption here is that multiple incremental versions need to be produced to aid in evaluating the final deliverable. Software development has been the classic use of this model and it has proven to be successful.
3. Quality—Models in this domain have emerged from the quality movement based on improved customer satisfaction that evolved from the 1970s. Six Sigma is the most mature of these current views.
4. Team management—Techniques that fall into this group often focus on the human side of the equation more than mechanical techniques. There are several well-known authors in this area but no dominant contributor. Detailed discussion on the HR-related delivery approach is outside the scope of this text, but should not be ignored in one's approach to improved delivery.

In addition to the model grouping above, the Project Management Institute (PMI) has been instrumental in documenting a broad view of required project management processes that are required across the full life cycle. Recent updates from this source have begun to recognize the legitimacy of a more flexible management view that is consistent with this text's goal. Beyond this, there is emerging evidence in the industry that excessive upfront planning is often not delivering the accurate specifications required in that model, which then raises the question of how to deal with that gap. The design goal outlined for this text is in agreement with this concept and will be represented in the upcoming version. One other well-recognized trend in this area is the survey quantified success of the iterative approach to software development. This is motivating interest in migrating that view into the mainstream of predictive-type projects, yet there are obvious mismatches in this view that need further examination.

Attempts to design an improved method for managing projects have stimulated the creation of many new tools and processes. This chapter will offer a high-level review of some of the most notable efforts and from this outline a taxonomy of their design structure. More details on the broader history of these models can be further reviewed by Alexander Moria's research as described in CIO (CIO, 2021). Nineteen specific models are summarized here but recognize that others exist. Categorizing the major design views for these models offers some insight into the underlying management goal. In other words, each model was designed to focus on some specific work characteristic. The following list has been loosely grouped by the underlying school of thought to highlight the fundamental design driver for that collection:

Predictive/Fixed Scope

■ Traditional waterfall—Microsoft Project is the most recognized software implantation of this model. Chapter 9 provides more historic and background details of this model.

- Critical Path—This term grew out of the early network scheduling model in the 1950s and remains a well-known concept today for sequenced tasks with fixed time estimates. The main functions of this model are to calculate task start and finish times, and the longest task path. This also means it is the shortest time to complete the project. The network concept has many planning and control uses.
- Dynamic systems development—The model focuses on alignment with strategic goals.
- Rational unified process (RUP)—An elegant modeling language created by well-known consultants. This model never achieved broad interest and remains a minority player now. It pioneered the use of schematic tools to describe requirements.

Iterative/Flexible Scope

- Spiral—Early approach to combing waterfall with iterative approaches; this is one of the first models to describe prototyping as an approach to defining requirements.
- Rapid application development (RAD)—This approach is essentially an implementation of the spiral concept. It predates the agile period and had an impact on that model specification.
- Agile—This model's defining principles are considered to be the modern school of the formal industry launch point. Chapter 9 explores this model in more detail.
- Scrum—This is currently the most used version of the iterative school.
- Kanban—This tool is imported from Japanese quality manufacturing lore and is now growing in popularity among iterative school dialects. It is used to show work throughput in a sprint and exhibits increased usage.
- Scrumban—Combines Scrum and Kanban for process improvement projects.
- Event Chain Methodology—Risk management process model.
- Extreme Programming (XP)—Designed to increase throughput by using short development cycles.
- Waterfall/agile hybrid—Multiple attempts are now visible in the literature to combine predictive with iterative concepts. All of these that were reviewed were considered to be niche approaches.
- Feature-driven development (FDD)—Requirements are based on defined features using small delivery teams.

Quality Management Oriented

- Six Sigma—The most mature and popular quality method in use today with a loyal industry following. Its roots come from the quality school this view has now been structured into a model-like process that is used in various small project development situations.

- Lean development—An analysis process designed to reduce waste (Toyota model).
- Lean Six Sigma—A heavily customer-focused approach to improving business effectiveness.

Team Development

- Crystal—A team-oriented model that focuses on project team interactions.
- Adaptive software development (ASD)—The focus is on team development.
- Team Software Process (TSP)—This team planning process is described in a 2000 Software Engineering Institute (SEI) research project to develop techniques for creating a high-performance team.

One thing that the variety of approaches included in this list shows is the lack of a high-level singular goal-based assumption view for delivering project results. If there was ever an example of the blind men and the elephant, this group represents it. For the most part, each model shown is heavily focused on only one element of the project as though that is sufficient to bring success. Each of these suffers in its scope of view and niche focus. They deal with only a small segment of the overall project architecture and none specifically recognizes the impact of a variable project profile issue as described in Chapter 3. Even though the profile variability is pretty clear as to impact it still leaves the question regarding how to integrate these into a broader-based coherent management approach. Personal experience suggests that most project environments tend to stick with a singular design view for managing their project suite.

As with most complex situations, the goal challenge here lies in constructing a model that properly guides the process through the life cycle steps and is first simple enough to understand and then match those steps to actual work requirements. Each of the models described above represents niche views of this process with widely scattered delivery techniques. From a high-level view, this grouping reveals how each seemed to be focused on a single gap target. In that regard, each of the models fell into the same trap as the predecessors. From a management scope viewpoint, they do not recognize certain macro-level aspects that affect successful delivery, and second, they look at the required work to be the same for all projects. Both these shortcomings will need to be resolved. Based on these observations, none of the historic models match the integrated design requirements stated but all have some interesting narrow perspectives that may have value. From this broad model overview, several gap attributes are recognized as being needed in a proper management design structure. These are:

1. The project profile must be part of the planning process and it will drive the subsequent work decisions.
2. Task execution must have the ability to use multiple work techniques within the same project structure.
3. The process must be driven by the delivery priority goals for selecting work strategies.
4. There must be a decision linkage between the strategic component of project definition to the ongoing status of selected projects.
5. Work management options must fit the project's deliverable goals.
6. Recognition of project resource management is required.

The six preliminary design items above have been cobbled together to represent a high-level review of the historical models and the recognition that the overall management scope needs to include the full life cycle of project decisions. The waterfall, iterative, and Six Sigma models are three examples of models that were initially designed to fit a narrow deliverable environment. Each of these is easy to understand from a goal standpoint and each has a task work structure that fits their design assumptions. They also have prescriptive views regarding how the work evolves through their different defined life cycles and each will produce the desired outcome if the project goal set fits that environment, but recognize that many projects do not fit them as we shall see in upcoming chapters.

To start developing an integrated model, one has to first understand what the proper decision process should look like. Even though this may be an overstated point, there is an appropriate key to map the project process. The first key step is to know what the output goals are and the second key is to understand the input characteristics of the project and its development environment. This view drives the success equation and the second key maps to the required work management processes that led to that conclusion. All projects have unique characteristics and environments on both ends of this and therefore these items must be part of the design architecture of the model. The historic review highlighted the four basic theoretical approaches to work design: predictive, iterative, quality, and team. Unfortunately, all four of these may be relevant in the same project but the following represents the essence of each category view.

Predictive—Known requirements are assumed; control is often a key driver
Iterative—Project requirements are less understood. This may lead to less formal planning and more team delegation; customer satisfaction is a key driver
Quality—Focus on customer satisfaction and less on schedule and budget
Team building—Creating a productive team is a universal management concept. This management area is relevant in the general equation but is left to be an internal management process within the external decision structure.

Examining this broad array of characteristics highlights why there are so many solution approaches. Managers are trying to satisfy all of these at the same time, yet the current model views only deal with one primary design approach.

In many ways, the job of a project manager is much like a medical doctor. When one goes to a doctor, the goal is the get the unidentified problem fixed quickly. There is an old vaudeville skit that goes something like this:

> *The patient says, "Doctor, doctor my arm hurts."*
> *The learned doctor says, "When does it hurt?"*
> *The patient responds with, "When I do this."*
> *To which the learned doctor's response is, "Then don't do that."*

We can learn a lot from vaudeville. A project manager must do more than this marginal doctor and look deeper to understand the factors driving the project life cycle to successful completion. This requires a deep understanding of the various knowledge areas involved as well as the dynamics observed in the ongoing project. Models such as these serve the role of providing a management decision structure to guide the related decisions.

Upcoming Chapters

Chapters 8 through 10 describe two classic predictive management models that have unique characteristics relevant to the new model design goal. Chapter 9 focuses on the iterative view. Chapters 11 and 12 examine two macro-level components that affect successful delivery. Chapter 13 gathers the macro-level factors that need to be in the new model. This chapter also begins the discussion by outlining how this model process can fit multiple project types. An initial integrated model decision block design emerges at this point. Chapter 14 describes the integrated model and how it will work across all project types. Chapter 15 describes the impact that the new model has on some key management processes. Recognizing that the model can look somewhat abstract, Chapter 16 offers a short tutorial for some of the key success-focused processes that will most change the traditional approach in those areas. Finally, Chapter 17 describes some of the backgrounds regarding how the model evolved from conversations about the industry with no initial forecast goal. This discourse outlined multiple iterative steps that led to the final structure.

Chapter 17 describes implementation issues that could be expected. No organizational change of this magnitude can be considered easy.

Reference

CIO. 2021. Why IT Projects Still Fail, March 3, 2021 (Accessed July 12, 2022).

Chapter 8

The Classic Predictive Model

Introduction

This chapter summarizes a more complete set of the internal model assumptions and then compares this to typical environment reality. The net result is a clear recognition that the model has many significant gaps with reality. Also, the model is not well understood by the typical project team and certain band-aid practices have worsened its value. On the one hand, the model is the most mature view of this environment, and on the other hand, it contains several bad management practices that make it of lesser value. Project failure is not credited so much to the model but the lack of understanding that its assumptions do not match the environment, particularly in the area of requirements definition. These gap issues are some of the primary target areas for the new model design.

The predictive model represents the classic view of project management. In this model, required tasks are identified, codified, and sequenced, which produces a clear deterministic schedule. The current collection of related theories, documented process descriptions, and long-term industry experience make this model require knowledge for all project managers. This model is simple to understand by all stakeholders and for that reason will be hard to change in most environments (a key point to remember).

The core portion of a working definition of this model is based on the concept that the project requirements can be defined and required tasks reasonably estimated, thus the outcome can be calculated. Historic roots of this classic and traditional method are traced into the late 1950s related to large government hardware (product) development projects. The model's view of the underlying cascading life

DOI: 10.1201/9781003431091-10

cycle task sequence process is similar to a real waterfall and that metaphoric view label used today is credited to Winston W. Royce in 1970.

Based on this view, the common industry name used for this view is the *Waterfall model*. Over the past 70 or so years many enhancements to the early simplistic view have been described by various sources, notably the U.S. DoD and Microsoft (Project software), plus large organizations and various consulting firms (Carstens and Richardson, 2020).

A popular way of viewing project schedules is based on Henty Gantt's early 1900 schematic bar diagram which is ubiquitous across most organizations. Figure 8.1 illustrates the cascading task waterfall view with schedules for the bars shown on the X-axis. The simplicity of this view shows why this chart format is so acceptable as a plan presentation view.

The view shown here illustrates the project phase view which is also a common life cycle perspective. The combination of Gantt charts and the sequential task steps are classic views of the traditional project. The four core defining artifacts of the predictive model are WBS, PERT time estimating, Gantt charts, and the network model (CPM). The lineage of these items provides two examples illustrating the level of resistance to change in this industry. First, once CPM as a concept became accepted the question became how to explain it. Drawing network diagrams failed. Attempts to draw project plans as networks did not fit the view that users wanted. A simpler Gantt chart became the replacement as the network task's view was reformatted to look like a Gantt bar. This section will elaborate more on how a model has unfolded since the 1940s.

As mentioned earlier, the DoD has been a major player in architecting the management concepts related to their large predictive project environment (DOD).

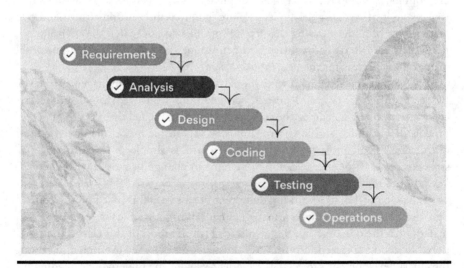

Figure 8.1   Waterfall project life cycle

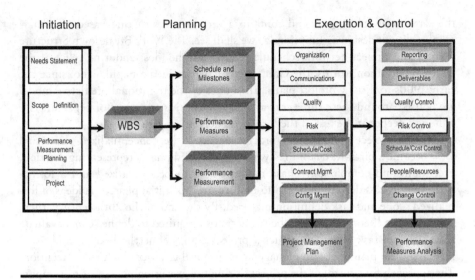

Figure 8.2   DoD project model schematic. Source: DoD C/SCSC,1976 (DoD)

Figure 8.2 shows a full-scale schematic of the formal process and management areas related to this model group.

This model has evolved since its inception and is now considered to be the de facto management view for projects, even though the iterative school is now increasingly recognized. This view is very much linked to the predictive model and is as close to a sponsored view of that model as exists. Note the full breadth of various support processes outlined in the structure.

Core Artifacts

Chapter 4 previously mentioned that PERT variable time task estimating as part of the late 1950s high-technology and high-risk environment. This model defined a statistical formula and theory to provide a method of evaluating schedule variability based on variable time estimates for tasks. At this early point (or even now) the user community was not mature enough to handle this more technically accurate tool, and task estimating practice lapsed to using a single discrete value. The second core tool, task networks, was not widely used for 20 years after its introduction, primarily because of the lack of computation support tools. As computer technology matured in the 1980s, networks re-emerged as the base schedule calculation mechanism. Both of these points are instructive in warning how hard it is to introduce new ideas into this community.

Much of the predictive model structure and process is familiar and most project managers would claim to understand it. Also, most industry research and documentation over the past 70 years has occurred around this basic structure. It represents the classic view of projects and contains a very logical structure based on a marginal reality fit design assumptions as we shall see. The Work Breakdown Structure (WBS) and project network schedule (CPM) are the most enduring relics of the 1950s. In addition to this, the Gantt chart has remained a popular tool since the early 1900s to display project plans. The bones of the traditional predictive model are housed around these three core artifacts. With this as a point of departure, we will look deeper at the rest of the formal model and its assumptions.

As the project management process began to be conceptualized the visual view became a birth-to-death life cycle with grouped phases representing creation, execution, and completion. Phase names began to be formalized and a formal planning process definition began to mature. In this initial phase, the ideas of formalized scope and task definition matured. By the end of the 20th century, there was extensive literature to describe the activities required to define scope, evaluate risk, estimate task effort, construct a project plan, define the budget, and use a formal project plan to obtain management approval to proceed with the execution. The approach for control of the project during the execution phase was to compare actual performance to the plan and use that variance to take corrective action. Assuming scope definition, risk assessment, and task estimating can be performed accurately this is an elegant model. The "beauty" of it lies in its ability to quantitatively predict a completion date and cost. Based on this, it also has sufficient detail to support management control for each task with defined start and finish dates. As an example, a predictive plan would show that Task 21 will be completed on July 5th, and the project will be completed on September 10th, with a total cost of $$800,000. This level of detail represents perfect visibility for management control! However, recall that Chapter 5 statistically described that project success rates typically do not approach 100%, and are often less than 50%. Project completion dates and budgets seldom if ever follow the developed plan. In essence, the predictive model is beautiful but generally does not offer a good match to reality. To assess issues with the model, management seems to want to have definitive answers even if they are wrong.

Predictive Model Assumptions

To assess this model further, it is necessary to more rigorously define what the model's underlying assumptions are and map those to a particular project profile. Technically, there is no formal detailed list of assumptions for this model. However, the description outlined here represents what most would agree to be the de facto definition of what the general industry defines it to be. The following list contains 11 implicit project design assumptions (note the italicized segments):

1. Project *delivery requirements and scope* can be reasonably defined before execution.

2. *Reasonable time estimates* can be produced for defined work units.
3. *Future risk events* can be properly evaluated in advance and proper measures taken to mitigate those items. A risk contingency reserve may be used.
4. *The task can be executed* according to the defined estimates and sequence.
5. *Project success* is measured by the planned time, cost, and deliverables.
6. *Timely quantity and quality resources* will be committed by the support organization according to the plan.
7. *Senior management and users* are readily available and standing by to help when problems emerge.
8. *Key stakeholders fully support* the project needs concerning technical and physical resources.
9. The project team members understand basic project management processes.
10. *Scope changes will be formally managed* and the plan adjusted based on approved changes.
11. The defined set of *user and technical documentation* will be produced.

Matching this list of assumptions to the typical project should make one question if the model represents the project environment. It is easy to see that the packaging of defined work units can be eroded pretty easily and the resulting plan can be significantly compromised for each assumption that is not met. This is exactly what happens in reality, yet the traditional project management process utilized continues, implicitly ignoring the fact that these assumptions are not completely valid. Without a doubt, the Achilles heel of the predictive model is the questionable ability to accurately predefine the project scope and related work units sufficient to accomplish the desired output. Secondly, a lack of ability to control the quality and planned quantity of technical resources will destroy even the best of plans. One approach used to mitigate these shortcomings is to pad the planned task estimates in hopes of covering both of the items above. Unfortunately, the padding process introduces its own set of errors which will be discussed later. Randomly adding values to estimates in the hopes that the actual result will come out that way is more like wishing than planning, so maybe the current traditional project plan should be called a Wish Plan instead because that is a better descriptor.

Most users of this model do not think about the level of assumption accuracy shown here, or what to do in situations where a particular assumption is not valid. An engineering definition of a valid model has *verisimilitude*—i.e., it matches reality. As an example, what happens if the project team is not competent and resources are not available to match the plan? One of the major values of any model is to aid in understanding a complex situation. If a project environment has this stated list of characteristics, the predictive model can't be questioned, but we have now painted a dark cloud over that view. One can argue that many of these assumptions are frequently suspect and generally are major gaps in many cases. The first three model assumptions are the most generally invalid—i.e., accurate initial scope definition is not the norm, discrete task estimates are not accurate, and risk cannot be reasonably defined. When gaps such as this occur, the model has to be patched and many of the patches used essentially destroy the integrity of model results.

The question now is if a project profile has major deviations from the assumptions list, what changes in the model are needed to improve the management of the project? It is easy to see that scope definition errors affect the related definition of work units and the resulting plan would be affected. Also, there are other outcome descriptive errors created by each assumption that is not met. This is what happens in reality, yet the project management process utilized plows on as though all assumptions are valid. Failure to recognize a false assumption is significant, but without a doubt, the Achilles heel of the predictive model is the questionable ability before execution to define the project scope and related work units necessary to accomplish the desired output. Unfortunately, band-aiding the model to fit reality in an attempt to overcome these shortcomings exacerbates the result as we will show later.

Role of the WBS

The predictive model uses the WBS to represent and define the project scope during the planning phase. This schematic diagram is one of the founding core tools and remains so today across a wide variety of uses and formats. A sample six-phase WBS is shown in Figure 8.3.

In this example, the boxes represent six project phases and the associated tasks under each phase box represent the work required to produce that phase of the project. Some tangible product-oriented projects will alternatively organize the WBS by parts (subsystems). This alternative view is called a product WBS but the task linkage logic is the same. Let's pause for a second here. The initial role of the WBS is essentially to aid in understanding the project, whether that be the phase view or a physical product component view (we'll see the product example below). In the author's view, the value of this artifact is in visually understanding the overall project, while the task linkage adds work details to that view. A second management value of the WBS is its numbering scheme. Note that both the major phases and

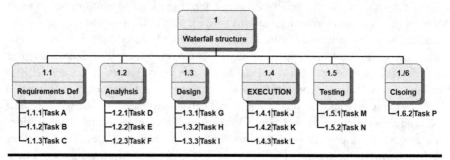

Figure 8.3   Sample life cycle WBS

underlying tasks all have code numbers linked for the phase (or subassembly). This hierarchical numbering scheme has value in work tracking and can be looked at as the project work mailbox. Many practitioners credit the WBS as being the most valuable management artifact in the project.

Another aspect of the WBS is its ability to show the evolution of the process through the life cycle. For instance, the first layer boxes might indicate how the overall project will be partitioned. Once that high-level decision is made, lower-level decisions follow downward as more details are added. The fancy name for this is *progressive elaboration*. Since the level of detail shown on a WBS is limited a companion support data source should supplement this. The industry called this a WBS Dictionary, but other names are used such as Project Notebook. Regardless of the name, this data repository captures the various decisions made related to the WBS boxes as they unfold.

Organizations and industries have diverse levels of formality and techniques for using the WBS but its general purpose is to structure the project scope, which then aids in defining a schedule and budget. Over the years, DoD has evolved rigorous standard WBS definitions for its family of major product groupings. The specification document for this is publicly available as Mil-Sts-881-E. This 291-page document is not for recreational reading but it does illustrate the importance that the DoD gives to this artifact. Each major product type (airplane, tank, ship, etc.) has its own formal WBS definition through usually three levels. As an example, the DoD product-oriented aircraft standard WBS template is shown in Figure 8.4.

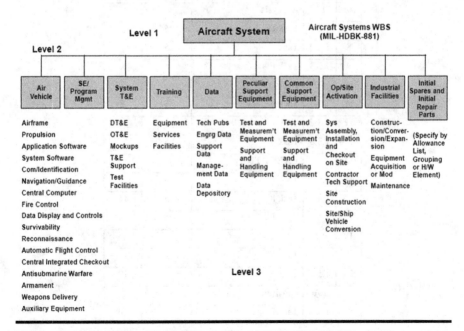

Figure 8.4   Aircraft standard WBS. Source: DoD Mil-Std-881

From the details shown here, it is obvious that the WBS can grow to a significant size in defining the project scope and the schematic view becomes unwieldy. An alternative format for this is then a table view, called a flattened WBS. Regardless of format, the use of a standard WBS approach serves two basic purposes. First, it standardizes how these items are and it also provides some cross-system analysis comparison for cost and schedule. The commercial usage of the WBS is not nearly so standardized, but the idea has merit. A WBS is considered the most common artifact for showing project structure.

As details are added to the WBS links can be defined between the WHAT view and move more toward adding the HOW perspective. For instance, once the required work task list is defined as being in the project there are five related data items needed to produce an initial project plan. These are WBS code, task name, duration (work days), cost, and sequence of task execution. As the level of detail grows it is necessary to convert the schematic structure shown in Figure 8.3 into a tabular format. An equivalent flattened tabular WBS task detail is shown in Figure 8.5.

id	WBS	Task	Duration	Link
1	1	**Waterfall structure**		
2	1.1	**Requirements Def**		
3	1.1.1	Task A	5	
4	1.1.2	Task B	5	3
5	1.1.3	Task C	5	4
6	1.2	**Analysis**		
7	1.2.1	Task D	5	5
8	1.2.2	Task E	5	7
9	1.2.3	Task F	5	8
10	1.3	**Design**		
11	1.3.1	Task G	5	9
12	1.3.2	Task H	5	10
13	1.3.3	Task I	5	11
14	1.4	**EXECUTION**		
15	1.4.1	Task J	5	13
16	1.4.2	Task K	5	15
17	1.4.3	Task L	5	16
18	1.5	**Testing**		
19	1.5.1	Task M	5	17
20	1.5.2	Task N	5	19
21	1./6	**Closing**		
22	1.6.2	Task P	5	20

Figure 8.5 Task details

This translation example is a little deeper into mechanics than a simple theory description requires but the importance of this step seems to justify that expansion. There are two critical data issues to explain. First, the cost field is omitted here but this data item would follow the same path as the task duration. Second, note that a line-item ID is added to the base data and its role is simply to provide a shorthand task linkage. The four core data items are:

- **WBS code**—This is the WBS reference ID from Figure 8.3.
- **Task**—This is the task name from the figure and is referenced by a WBS code.
- **Duration**—This is the estimated work time to produce the task (usually days). In this simple example, all durations are assumed to be five days.
- **Link**(predecessor)—This is the most complicated item on the list. The numbers in this column are related to the order of execution. Task D has a link code of 5. That means it follows ID row 5 (Task C) in the work sequence. These codes are used to simplify the linkage information. Technically this column is called the predecessor link list—i.e., Task C is the predecessor to Task D.

This example represents the level of task planning detail required. From this data specification, computer software can mechanically generate an equivalent network plan and convert that view into a visual bar Gantt chart. One can see the compelling nature of this process as the plan appears automatically from the software model.

CPM Networks

The second step in this expansion is the role of a project network. As indicated above the introduction and use of task networks is one of the 1950s core tools of traditional project management. The mechanical role of a network is to use task estimates and sequencing to calculate the project schedule. Chapter 4 described the origin of the Critical Path Method (CPM) role as a scheduling tool. Figure 8.6 shows a simple demonstration of the underlying network that is used to compute the longest path. This is called the AOA format for Activity-on-Arrow, meaning that the tasks are represented on the arrows with start/stop indicators as numbered nodes. Also, note that the software handles all of the calculation complexity using the five data items above. The solid arrows represent the calculated critical path through the task list and scheduled completion dates are shown for each node (i.e., start and finish for each task).

The sample CPM view shown here is not based on the earlier WBS sample project, rather it is selected here to better show the task-level linkage architecture. The sample program model in Figure 8.4 is too complex for an introductory example. It is also important to recognize that there are two network camps. One group wants to see the tasks as arrows (Activity on Arrow) and the other camp wants to see the

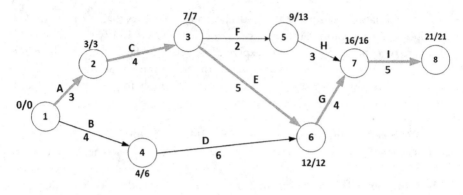

CP = A-C-E-G-1; And 21 time units

Figure 8.6   Project AOA network model

same with tasks represented as nodes (Activity on Node). There are minor differences in mechanics but the goal for both is the same and that is to compute the time required to complete the project along with identifying the longest path through the network, called the Critical Path. The network in Figure 8.6 uses arrows to represent tasks. And nodes simply represent the start and stop points. As an example, if task A was five days and task C was five days, the project should arrive at node 3 in ten days. Calculating total network time becomes more difficult as the size of the network grows and parallel paths enter the process. The calculated schedule times for each node are represented by E/L values. For example, at node 5 the value is 9/13, which means the earliest time this can be reached is 9-time units and the latest is 13, so there are 4 units of slack or extra time in that path. Those nodes with no extra time are on the critical path, meaning no extra time.

As one can see, project networks with more than 25 tasks become unwieldy to manually calculate, so it is easy to understand why this technique was in limbo until computational support tools became available and brought this idea back to life in the late 1980s. The use of CPM networks highlighted the role of task linkages (predecessors) in computing a schedule. Microsoft Project and other similar software packages have made this process a universal method for schedule and budget planning.

Initial Project Plan

With the previous two planning core tools explained, we can now begin to see why those predecessor explanations were required. To add to this explanation, it should be obvious that the amount of task-related data involved in planning a project can get overwhelming and the mechanics to calculate the network parameters even worse. Historically, the role of the network was understood early on but the operational maturity of this idea lingered until at least the 1980s when computer hardware and software began to catch up to the problem. Performing this calculation is the epitome of the predictive model—i.e., everything is now quantitatively predicted.

At this stage, the role of a WBS and network logic became foundations in the model approach to planning and control. Little attention was paid to the error gap related to scope definition and time estimating. We can now use the task estimating data shown in Figure 8.5 and process it through a software tool (such as Microsoft Project) to automatically produce an initial schedule. More elements may need to be added to this view from follow on decisions but this initial view represents the core defined work activities from the WBS. In a real project, the volume of data grows considerably but the same mechanics can be applied as shown in this example. Figure 8.7 (later in chapter) shows the results generated by the calculation software (using data directly from Figure 8.5). Task schedule information is not shown here but is calculated by the process and is available. The important concept represented by this example is that the plan produced is directly linked to the scope definition and therefore has some validity so long as the parameters are correct.

There is nothing magic about this calculation process but it is a very handy aid to the planning and control process. Here are a few points to raise from the plan shown:

- The plan table values are the same as those developed from the WBS, which adds integrity to the plan if the scope definition is correct.
- A project starting date can be used to move the scheduled starting point.
- The software schedule obeys a work calendar so weekends do not count on scheduled task calculations.
- Various data can be placed on both the table and bar view. Note that Task P shows a completion date of 10/28. Similar data can be shown for all tasks.
- WBS phases are preserved for reference and task grouping (heavy black bars).
- Note that the output format looks more like a Gantt chart than a network; however, this view duplicates the network by using bars with arrow linkages. Recall that this is the way the industry wanted to see a plan so the underlying network format was converted to look like a Gantt chart with bars.

The Internet contains many sources to illustrate the keystrokes required to produce this view. This is a simple application of the software, but it is clear that much of the future management process is moving into more complex uses of information technology for all phases of the life cycle.

WBS Listing with Plan Parameters

The discussion and examples shown thus far have been model theory-based. This has not meant to say that all project plans are produced this way, but that is the model approach. The method shown here forces a strong link between the planned project elements and the resulting defined work tasks associated with those elements. The best practice for translating a WBS view into a project plan is to reference all project work elements with an appropriate WBS code as illustrated by the previous example. By doing this, all project work is linked to a defined deliverable. Carstens and Richardson offer a more detailed description of the WBS and network role in the planning and control processes (Carstens and Richardson, 2020).

The model artifacts shown here represent the core of the traditional theory. Most project managers have used a similar tool approach for planning and scheduling. Regardless of the underlying process used, the Gantt chart output plan format is a very common view. Given the long history of looking at projects in this manner, this classic view will be hard to change despite obvious gaps in reality.

The initial project plan has illustrated how the software hides the network and displays a Gantt-looking chart. One must now recognize that the predictive model is simple to understand by all concerned and for that reason, many users will be resistant to change their approach (a key point to remember). If one looks back at the initial plan figure, recall that the scheduled completion was calculated to be 10/28. However, what happens if one of the tasks overruns by ten days? Mechanically, the plan should now show the date to move by ten days but it does not. Recognize at the outset that there is essentially no real expectation that the stated completion date is valid, yet it is often quoted at the planned date. An old professor once said that one should never express a data value that did not reflect accuracy. In this case, the model is going to be wrong for more reasons than has been explained thus far. In reality, the dates shown are targets and various schemes are used to make the equation come out ok. Most of these patches destroy the integrity of the presentation and create poor operational practices. One of the common techniques to make the dates and budgets come out correct is to simply add padding to each task in hopes that the result will be close to that. In the author's opinion, padding tasks as described turns the resulting plan into a wish plan with no substance to it. In that form, it does not represent valid management. More on this is coming after we cover the issue of scope changes.

Scope Changes

As indicated earlier, the Achilles heel of the predictive model is scope change after the plan is approved. Review Figure 8.7 and examine the impact of adding new tasks to the project as a result of approved scope changes. It is easy to see how this action can expand both the schedule and budget. More importantly, this is a core area that relates most to operational management issues. The rate of scope change on a project can easily be 2% per month. If not handled properly, that level of change can increase the schedule and budget by around 25% per year. It must be recognized that this is now a different project and the active plan should reflect that. Scope changes are not an overrun as often recorded based on a fixed

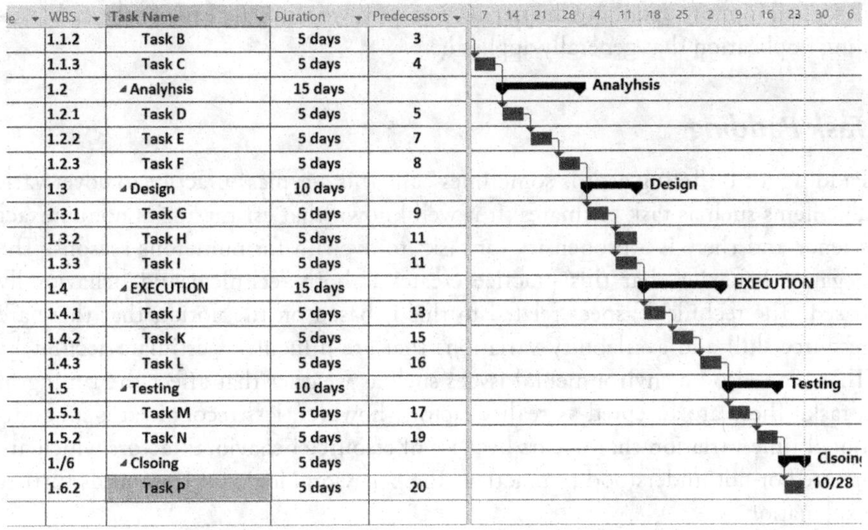

le	WBS	Task Name	Duration	Predecessors	7 14 21 28 4 11 18 25 2 9 16 23 30 6
	1.1.2	Task B	5 days	3	
	1.1.3	Task C	5 days	4	
	1.2	◢ Analyhsis	15 days		Analyhsis
	1.2.1	Task D	5 days	5	
	1.2.2	Task E	5 days	7	
	1.2.3	Task F	5 days	8	
	1.3	◢ Design	10 days		Design
	1.3.1	Task G	5 days	9	
	1.3.2	Task H	5 days	11	
	1.3.3	Task I	5 days	11	
	1.4	◢ EXECUTION	15 days		EXECUTION
	1.4.1	Task J	5 days	13	
	1.4.2	Task K	5 days	15	
	1.4.3	Task L	5 days	16	
	1.5	◢ Testing	10 days		Testing
	1.5.1	Task M	5 days	17	
	1.5.2	Task N	5 days	19	
	1./6	◢ Clsoing	5 days		Clsoin;
	1.6.2	Task P	5 days	20	10/28

Figure 8.7   Simple project plan

plan. An earlier chapter described the use of a scope reserve for changes. This is the proper way to recognize this activity. We can now see how the use of the reserve can protect the overall plan without relying on padding the tasks blindly. If the reserve is 100 units and a scope change requires 10 units, that amount is extracted from the reserve and added to the plan with the new tasks. This leaves the overall plan with reserves intact. Technically, scope changes create not only new tasks and time increases but additional budget as well. For large changes, it may be administratively worthy of adding both additional budget and schedule to the approved plan but in many cases, a change is so small in duration that it may not be worth the administrative time to reflect so only the additional budget is tracked and the planned schedule remains static. This event is a quandary where theory meets reality. Adding both schedule and budget from the scope reserve to the plan is an issue that must be recognized. There is no easy answer to resolve how best to manage the small changes, but the ironclad rule is that no changes will be made unless formally approved and a scope reserve will be allocated to support these actions.

Task Estimating

One of the worst management planning practices is padding task estimates to cover potential variability. On the surface, this is a perfectly logical idea but most do not understand the psychological-based negative behavior that this creates. The section below comes with credit to Eliyahu Goldratt's Theory of Constraints which will be described in more detail in Chapter 10. For now, we will only explore the task padding implication that generally applies here.

Task Padding

Padding (or buffering as it is sometimes called) is a typical practice to cover variable items such as task estimates. It is well known that estimating is not an exact science and there is a propensity for tasks to overrun for numerous reasons. The negative behavior that this practice creates is both technical and behaviorally based. The technical aspect related to this is based on the notion that there are resource skill and availability variations that are difficult to estimate accurately. There can also be environmental issues such as weather that affect the timing of a task. These are accepted as reality factors; however, less recognized is an additional time variation that occurs because of resource behavioral factors which are ignored or not understood in practice. The behavioral logic portion needs further explanation.

Behavioral Variations

One way to understand the behavioral side of task time variation is to look at your personal history. Let's go back to our school days for a likely personal case example. When you had a five-hour school assignment given on Monday and due the next Monday (seven days later), when did you start working on it? You thought you could execute it in one day and there were other things you needed or wanted to be doing. My strategy was always to wait until Sunday evening but you may be more disciplined than most. There is a human propensity to delay working on things for various reasons (some valid and some not so much). This behavior is so well known that it is called the *student syndrome* and it often stays with us all through our lives. The related behavioral trait described is *procrastination*. This trait certainly is alive and well in the project world. Here is a more specific project-related description of this phenomenon. A task has been padded by 100% to cover potential overruns, and this is not an atypical padding level. The project team knows the level of padding, so the tendency is to procrastinate starting the task and use up all of the padding. This delay is assumed to not cause an overrun given the padding. Later, when it comes time to execute the task, and all the padding is now gone, the same result occurs as happened in the earlier school scenario—a late-night Sunday crisis to finish on the Monday schedule. In the project case, some factors can cause the task to take longer than the raw estimate, so the net effect of task padding as described here is that there will still be time overruns despite the logic used to protect this event. Understanding this padding scenario is one scheduling lesson in project management one must learn and adapt practices accordingly, Padding does not work! The Critical Chain model has rigorous rules to deal with this phenomenon. There is more detail about this in Chapter 10.

Risk Events

Handling project risk is rated as the most immature of all project management processes. Much of this appears to be caused by project managers who do not believe that they can anticipate such events and thus give the process a short change. Recognize that there are formal risk management models in existence and if these work for your project that would be good, but none of these are perfect predictors. Previous sections of the text have described the potentially significant impact that an unplanned risk event can have on the project, either during execution or later after production. *Some level of risk assessment on every project is mandatory.* Do your best within early planning time constraints and hope that this effort has been reasonable preparation for what is to come. In many cases, this assessment process is not glowing in accuracy. Regardless of the situation, handling the risk inherent in the project life cycle is a challenge. The risk reserve concept outlined earlier provides the mechanics to handle the events once they are visible. Research studies indicate that the identification process is not overtly accurate, especially early on in the learning curve. However, one factor did emerge that is important to add to this equation. That is something called *risk culture*. Humans are pretty good at anticipating risky events when trained and motivated. This means in the project case that it is important to pursue training on the impact of risk events on successful completion. Efforts should be made to identify and mitigate the items defined. From a theory standpoint, there is extensive risk management literature available and this is a learning area by itself beyond the scope of this discussion. Also, it is good to learn about the impact of risk events on other similar projects and some of the causal factors. One simply has to believe that this is a critical success factor and there are methods to improve the identification and handling of this class of events. The key to managing risk is to understand that these have not yet occurred, and therefore they are not in the core work plan. Resources for handling these events are extracted from a risk reserve. Once triggered, the additional work is moved into the action plan along with the additional resources from the contingency reserve.

Now, a final management point on project risk. Using the WBS as a directory of task scope, assign risk oversight responsibility to individuals who understand that area best. These individuals are called *risk owners* and should be formally identified by a WBS code or role designators such as fire wardens or other role titles. These individuals become the front line of defense in quickly identifying and reacting to these events as they emerge.

One way to look at risk is to think of it as anything unplanned that affects the project. In that view, project management becomes risk management (or vice versa). If the project behaved according to the published formal plan, the role of project management would be task completion checker and no one has that view of project management.

Project Control

Although there may be a counterargument to this statement, the author judges one of the primary underlying goals for using the predictive model is to control the outcome. It is not to optimize time or user satisfaction. The predictive model mechanics offer specific management control status parameters for each task and the total project. Laying on top of all of this is the defined process to monitor actual results versus the plan. The term "plan versus actual" is a common statement of status. To perform this comparison, the model approach is to "freeze" or "baseline" the plan after post-planning management approval, and these data values are used to measure performance. This is another messy topic related to the model. We have already seen the impact of scope change and risk on a fixed plan. If the reserves described earlier are not used as outlined, the project often becomes a numbers game between the project team via padding and the external review sources. One way of hiding the actual status is to simply add padding to the tasks and measure the plan versus the actual base using the padded value. This is not control! The goal of project management should be to openly communicate the truth about the project. Management has historically caused a problem with this by viewing status negatively when the comparisons showed variances. In this situation, the project team's performance is often blamed when we have easily shown that an overrun can come from numerous sources. The appropriate control goal should be to produce an honest plan with appropriate reserves and monitor variances so that corrections and improvements can be made. In other words, the concept of project control should be focused on learning and correcting more than seeking blame. Project managers are not dumb and they are much more knowledgeable regarding the workings of the project than senior management or stakeholders. A personal opinion is that too much of the predictive model is focused on false predictions without working to understand what is happening concerning factors such as scope change, risk, and task variances. To examine this situation in another form, a discussion of the agile model follows this chapter. Note the different views of control found there. In the agile case, control is heavily focused on the internal project team's work status and much less on higher-level management variables related to time and cost variances.

Management Reserve

The sections above have described project variability resulting from factors essentially external to the defined task work estimates. We have reviewed the impact on the initial plan that scope change triggered by new insights into the problem. Similar changes can come from risk items not recognized in the initial plan. In both of these scenarios, two defined reserves are recommended to fund these two common external plan variations. Chapter 6 described how variability in plan entities should be handled via multiple reserves. The internal task variability is handled within the plan by buffers attached to the core work plan, when task variation occurs the overrun is protected by a buffer which we will see in more detail later.

The previous discussion has questioned task planning accuracy. Now we have to recognize that a task's duration can vary for numerous other reasons—i.e., weather, resource availability, bad estimate, etc. It is desirable to know which factor created the overrun, but most are not willing to go that far with task analysis and evaluation. The recommended option for this event is to establish a management reserve for the plan to cover task variability. This reserve is designed to cover both schedule and cost overruns caused by work task variations. Tactically, this can be shown as a project buffer. Figure 8.8 shows a project buffer (1.6.1), which is added to the plan completion date originally shown as the base plan.

Note the buffer is sized at 14 days and this causes the new project schedule to be expanded by that amount to "11/10." The buffer is designed to cover task overruns outside of scope and risk factors. The formal name for this is management reserve but the operational name is often called a *buffer*. The use of buffers in a plan is often looked at by management as padding but as described here they are defined by logical techniques to cover operational variations. Some organizations have complex formal approaches for dealing with task overruns, but the basic logic should be similar to this view. The mantra here is to use honest task estimates and cover the project variance with a named reserve.

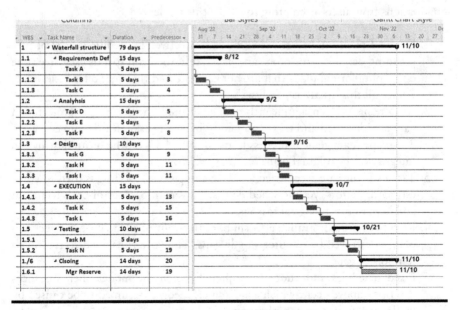

Figure 8.8   Project plan with management reserve buffer

Control Accounts

A control account can be arbitrarily defined in the WBS by grouping tasks into manageable groupings. This grouped view of the project work is handy for both cost and work management reasons. In the case of cost management, a defined Control Account provides the structure for actual cost collection (see Richardson and Jackson, 2019, for more on this topic). For the theme of this text, the more important related idea is the use of a *Control Account Manager* (CAM). The use of CAMs is not a popular technique but one that needs to be considered. If one looks at the Control Account as something that needs to be managed, then it is worthy of formally defining a CAM to oversee that segment of the project. This delegation-type approach has the potential to improve team buy-in and can also be a morale boost for the team if decision-making authority is passed to the CAM.

Predictive Model Reality Match

When examined through a critical reality lens the predictive/waterfall model does not represent the normal project environment very well in many cases, and even less so if not managed properly as outlined here. However, this should not necessarily lead to the conclusion that the model is completely invalid, but rather emphasize where those gaps exist when a specific assumption does not fit the particular project profile and then make appropriate adjustments. The following are typical sample observations from users attempting to follow this model:

1. Detailed planning before execution is ineffective. The required level of pre-planning detail is error-prone and as a result, the team spends excessive time developing formal plans that are not accurate and therefore offer little value. Some call this "analysis paralysis."
2. Requirements often significantly change through the life cycle as project plan gaps are identified and this causes additional confusion for the project team. The scope management process leads to plan changes which can destroy the integrity of the original plan.
3. Task estimates are recognized as being complex, so padding is used to cover estimating errors, yet the schedule is still overrun because of both technical and human psychological reasons.
4. Resources are not available to support the defined plan. Even an accurate plan is destroyed by resource supply gaps. Time overruns are directly linked to resource gaps.
5. Fixed schedules and budgets are predefined in the model, then variances described above occur which are often erroneously assumed to be caused by ineffective project team performance.

6. Project teams often focus more on meeting the predefined task schedule and less on the actual needs of the users (i.e., Is the actual priority goal of the effort completion time or customer satisfaction?).
7. One of the major model constructs that are wrong is the concept of a single-task predecessor. This view looks at the schedule unnecessarily. For example, if two rooms need to be constructed and then painted, the model might define that they are constructed first and then painted. What about the alternative of finishing each room? This can indeed be coded if the human sees the option but ideally, the model should evaluate both options more openly.

If one accepts this description of predictive model reality gaps, there is certainly motivation to look at techniques to improve the operational integrity of the approach utilized and related management process. On a more positive slant, there is a great deal of validity in the way the predictive model describes the work required to produce project deliverables through its mature defined life cycle management processes.

Industry performance statistics have long pointed out that something needs to be done to improve project outcomes. There are two very fundamental management issues needed to improve the waterfall approach:

1. More effective processes to deal with handling scope changes during execution. Freezing the approved plan is typically not the right answer, yet changes erode the approved plan and create variances that negatively impact schedule and cost forecast results.
2. Poor resource management invalidates any plan. More will be discussed on this topic in a later section but for now, realize that having no timely resources applied to a task when needed will likely result in a schedule overrun. That sounds pretty simple but this variable is easy to observe in real projects as a common occurrence.
3. If an organization is to improve, it must define methods that did not work and execute actions to improve those items on the next project. This is the learning organization concept. Hiding problems and not reviewing problems limits that ability.

The predictive/waterfall model is so ingrained in the project management history that much of it will remain for the foreseeable future. What is important for this discussion is to recognize why the model has known gaps in matching the reality of a project. The question remains at this point to define what can be modified to have a better match. Identifying techniques to do this is the challenge faced here. It is important to recognize these gaps and not just follow the model blindly.

References

DoD C/SCSC. 2018. Cost/Schedule Planning Control System (C/SPCS) Specification. https://www.acq.osd.mil/asda/jrac/docs/DoD-Directive-5000.01.pdf (Accessed August 16, 2022).

Carstens, Deborah and Gary Richardson. 2020. *Project Management Tools and Techniques.* 2nd ed. Boca Raton, FL: CRC Press.

Richardson, Gary and Brad Jackson. 2019. *Project Management Theory and Practice.* 3rd ed. Boca Raton, FL: CRC Press.

Chapter 9

The Iterative Development Model

A Changing Mindset

For much of the current history of project management, there has been a wide variety of organizational groups defining methods to manage the execution of a project (i.e., DoD, PMI, and various large organizations). Much of this previous model development has been oriented toward satisfying management entities with specified data related to planning, control, or tracking status. As a result of this focus, the models often focused on predicting the final schedule and cost, even before the project is approved. Also, during execution, the traditional tracking process compares planned versus actual values to ascertain status. As a result of this cultural control mindset, the traditional view of project management is oriented toward this view. As an example, the waterfall model defines the start and finish date for each task as well as the cost of that work unit. All of this plan detail is produced from a defined project scope and related work estimates. Chapter 8 showed examples of the classic predictive environment and scribed how the plans produced from this effort do not fit reality for a wide variety of reasons, particularly in the accuracy of predefined requirements. Recognition of this accuracy problem has raised the question as to whether that level of planning effort is worthwhile.

In some tangential way, this dissatisfaction with the traditional model may well have been the root stimulus for a new approach. The emerging popularity of the iterative school of development fits that belief. In the latter 1990s, active work was evident in this direction. *Prototyping* was an early term for this, but the approach eventually became known by the more current term *iteration*. The most successful strategy from this era is now called the agile school of management.

DOI: 10.1201/9781003431091-11

Agile History and Trends

Nothing has changed the project management landscape over the past 20 years more than the concepts surrounding the terms "agile" and "sprints." This historical view offers insights into the related delivery methods that have proven successful in changing organizational project management culture in a relatively short period.

One of the first widely recognized using the iteration idea for development is credited to Barry Boehm in his 1986 case study of a large software development project using what would now be called *iterative prototyping*. The title for this early classic model approach was "Spiral," based on the metaphor that the answer was spiraling closer and closer to the desired ending as new versions were reviewed. Later versions of this early delivery method were derived but never were accepted for various reasons, likely related to a frozen organizational bias against the loose control approach. The idea of prototyping software requirements remained of interest in that industry segment, but still not in the mainstream of "legitimate" methods. Many of the early efforts related to this approach seemed to be more of a motivation for ways to avoid the time-consuming documentation bureaucracy required by the predictive methodologies. Formulation of the method into an acceptable model format had to await another initiative that we are calling the "agile revolution."

History of Agile

The current view of iteration as a delivery model occurred in 2001 when 17 software practitioners who had been working with this approach for software development met at a ski lodge in Utah to formulate a single view of the process. The outcome of that meeting was a vision document titled *The Agile Manifesto,* which is now recognized as the defining vision document for the method. In addition to this high-level vision statement, the group defined the following 12 supporting principles designed to provide more specific guidance for the new management approach (Agile Manifesto):

1. The highest priority is to satisfy the customer through the early and continuous delivery of valuable software.
2. Welcome changing requirements, even late in development. Agile processes harness change for the customer's competitive advantage.
3. Deliver working software frequently, from a couple of weeks to a couple of months, with a preference for a shorter timescale.
4. Business people and developers must work together daily throughout the project.
5. Build projects around motivated individuals. Provide the proper environment and support their needs, then trust them to get the job done.
6. The most efficient and effective method of conveying information to and within a development team is face-to-face conversation.

7. Working software is the primary measure of progress.
8. The sponsors, developers, and users should be able to maintain a constant pace indefinitely.
9. Continuous attention to technical excellence and good design enhances agility.
10. Simplicity—the art of maximizing the amount of work not done—is essential.
11. The best architectures, requirements, and designs emerge from self-organizing teams.
12. At regular intervals, the team reflects on how to become more effective, then tunes and adjusts its behavior accordingly.

As evident from the list, the target model was originally focused on software development. Note that the general flavor of the principles is more of a delegated one at the team level, with less focus on higher-level management control. Early experience with the model spawned dialects with minor process changes to the original somewhat bland approach. The general result of this was significantly improved user satisfaction with the outcome. However, pundits of this view claimed that this experience did not reflect the real world of project development in that the projects were relatively small and there was less focus on completion schedules and budgets. Nevertheless, industry survey results compared the method to traditional projects and the results favored iteration over traditional predictive methods (see Chapter 5). Formal evidence of this type and published case studies stimulated an increasing interest in expanding the use of iteration techniques in traditional project environments. The one primary blocking constraint for broad project-type usage is the belief that tangible deliverables do not fit well with a prototyping approach. Despite this limitation, the positive project results with this approach created a movement to find more ways to follow this direction.

Note that the formal introduction of agile was spawned from a grassroots movement among working-level technical types and not from the more typical large host organization such as DoD or PMI. Instead, the agile group consisted of technical specialists who understood that part of the life cycle and did not feel a need for management help in the process. At its design core, the delivery unit is a focused work area housed in a fixed timebox. This work process is called a sprint which is designed to produce interim deliverable segments using short work cycles. The actual physical characteristic of the output from this work process was only known when the sprint was completed. A user evaluation of the output would be the key to deciding on the status and the need for further sprints to improve the outcome. Depending on one's bias, this perspective is either good news or bad news. Regardless, the argument continued that this only works in environments where the deliverable can be decomposed into chunks of usable output. In other words, it is only applicable to software development.

One of the new ideas introduced in this development approach is to view the required output loosely in terms of a deliverable "feature," rather than some fixed definition (think product blueprint in the traditional specification jargon). A sprint is assigned some collection of desired output features to work on and the team does that in the time assigned, typically in two to four weeks. An evaluation of the results determines the status of the feature and plans are made accordingly to execute another sprint or assume the product is finished. In reality, there is some higher-level constraint on time and budget but that is not the primary focus. This process continues until the customer is satisfied, while the traditional complaint is that there is no control over this process.

One of the more visible aspects of this process is the active involvement of users which is also recognized as a project failure factor in the predictive model. Second, there is much more deliverable movement forward here as preplanning is much shorter and delivery "chunks" are available sooner. The long delivery cycles of the predictive approach are known for customer complaints. There is little argument that these iterative model attributes represent positive virtues, but there is still a bias that says there is more to a project than this. In other words, there is a higher-level perspective that is not part of the agile view. One legitimate view of this question is "how does the organization know that this project was the most important one compared to many others that could have been selected?" There are other similar traditional management issues related to risk, status tracking, and general project control. One has to look at the higher-level issues in more detail to understand this bias. More of this will become visible as future ideas are described. For now, the goal is to understand the value and mechanics of the iteration process.

By 2010, acceptance of the iteration method had progressed to be the de facto method for software development and large organizations were adapting to that view with various incremental management processes. At this stage of evolution, the arguments between iteration and predictive schools of thought falls into religious camps. As the method continued to invade predictive environments, the traditional development culture continued to express opinions that the new method was not appropriate for a tangible product that can't be produced by multiple iterations. Also, there is a bias that judges the model to be appropriate only for small, low-risk, and low-complexity project types. Both sides of this disagreement have merit in their views but the goal here is not to take sides on this unresolvable question. Maybe they are both right and wrong! Dealing with this further will be the goal later in the text.

One of the fuzzy philosophical views underlying the agile approach is the belief that a project-predefined task management plan has an error level sufficient to make it worthless. From that scenario, much of the planning cycle creates documents that have little work value. The agile school answer to this is a two-edged sword. Is the goal to get something done quickly, or spend some time deciding on the target to be pursued? Some forecast estimate of deliverable variables is needed for project selection but this does not have to be in specific detail. The iterative approach suggests that the final answer will be better defined from a user standpoint. Regarding some of the more traditional environmental tasks, one will seldom notice a project team is excited by the opportunity to produce any of the formal documents outlined in the traditional project methodology—i.e., Charter, Project Plan, status report, user guide, a technical guide, etc. Much of this type of output is absent or minimized in the iterative school views and this is yet another area of conflict. Most organizations require various degrees of this class of project output. Where does it fit in the iterative project plan? Before agile methods will be widely accepted in the traditional organizational world, these philosophical planning, control, status-tracking, and supplemental output roles will have to be formally resolved. If the organization's management structure has confidence in the project team, there would likely be decreased resistance to approving a less detailed level of status information. Accepting the iterative approach as a positive could lean those organizations more toward the concept of less formal requirements definition and other iterative streamlined techniques. The goal at this point is not to answer this question but to highlight that this is a potential work execution method that is in transition from a cultural view and one that has a high probability of being increasingly accepted in some environments. In many ways, the iterative evolutionary approach to project delivery has followed a similar wandering leaderless path of other management methodology techniques.

Iterative methods have now reached the maturity stage where they are viewed more positively, although the approach still does not have the full perspective of the project life cycle. The key positive attribute that remains as a core driver involves its proven role in achieving high user satisfaction but remaining resistant to its lack of visible planning and control. One thought regarding how to approach a resolution to this would be for senior management to "soften" the level of traditional status tracking required recognizing that this is not adding value to the effort. This would allow the project team more time to focus on delivering the stated features but with less oversight. That change in management culture may well be a hard sell in many organizations. One very legitimate question regarding the less up-front planning concept remains whether it is viable in an environment where formal schedule and budget data are required before a project is approved. Reducing the degree of status tracking is counter to traditional views. Individuals who are biased in the iterative direction would argue that the sprint process does produce adequate status tracking, but that argument also remains unresolved.

The Traditional Tracking Myth

One of the bastions of traditional project management is the practice of requiring a formal work plan that outlines in detail the planned outputs, schedule, and budget. Never mind that this document seldom if ever equates to the actual project results. Scope changes, risk events, task overruns, and other environmental events cause the plan to typically be in error even with heavily padded time and budget estimates.

In many ways, the traditional approach to planning is more like trying to guess the future answer than quantifying the actual known requirements to produce the plan. This use of the traditional detailed plan data is not evident in the iterative process. On one side, the organization needs project parameter forecast data for aggregate planning and some project environments need firmer control of the planned deliverables. As with most complex situations, the correct answer could well be "it depends!" Neither zero planning nor excessive planning is the correct answer. This topic is certainly a candidate for future research and long-term discussion. Through all of these divergent views, one can see both sides of the argument. In any case, a management model should allow local bias to be used and not dictate an assumed view. That is the mantra that will be followed here.

One of the success attributes verified in the agile methodology is the active role of future users in the execution process. This is often not the case with traditional projects as quantified in various industry surveys outlined in Chapter 5. This is a known best practices requirement, but agile somehow has been able to make it more of a reality. Traditional projects would be more successful with the same strategy. This type of involvement helps to quickly correct errors that might not be found until later in the life cycle, maybe even after implementation.

One of the interesting historical events related to agile that we will come back to later is how it was spawned from a knowledgeable working class of technicians who sought improved technical solutions to their work process and not from some high-level management perspective. One might conclude that the creation of a development method would reflect how the founder envisions the problem. This simple observation may well have a long-term influence on how this issue evolves. A successful resolution is going to have to deal with these different perspectives with the character of the result emerging in the middle of that. The following list contains perspectives that influence different various outcomes:

1. User satisfaction seems to be higher with fewer front requirements specifications.
2. Projects compete for the organizational resource so forecast data is needed.
3. Project failure is observed due to inadequate initial risk assessment.
4. Management feels responsible for the global status tracking of projects.
5. Several project-related asks are more fixed in format than iterative.
6. The iterative project domain view does not cover the full project life cycle activities.

In reviewing this list of diverse perspectives, one might conclude that the proper view of iteration is to be embedded inside a broader model that deals with the other management aspects outlined. At any rate, this breadth of management and technical segments needs to be addressed. The one irrefutable conclusion regarding the iterative approach is that the user is more satisfied with the result than with the process related to predefining the requirement and then uses a formal scope control process to initiate changes. Also, the ineffective use of a formal plan as a control tool makes that process a candidate for change. From these various diverse points of view, we see both optimism and immaturity in the iterative approach. It has certainly found a loyal following and the quantitative evidence supports consideration of it as a delivery model.

Agile Dialects

As evidence of process immaturity and lack of central guidance, the original agile development vague approach produced a significant variety of new versions each touting improvements in the basic concept. At this stage, there are three major emerging versions based on the agile iterative principles. These are classic agile, Scrum, and Kanban. Within this collection of dialects, the following benefits are touted (Distant Lab):

- Better Quality Products
- User Satisfaction
- Enhanced Control
- Better Product Predictability
- Improved Flexibility
- Continuous Improvement

Detailed mechanics related to these three dialects is beyond the scope of this text, but the concept of loose requirements definition and team sprints is common across the groups.

Once the agile process was demonstrated to be successful, dialects of the method began to surface with each claiming to be better than the previous. Various surveys offer the approximate usage frequency mix as follows:

Agile—4%
Scrum—58%
Scrumban—10%
Extreme Programming (XP)—8%
Hybrid—9%
Kanban—7%

Assuming these data are approximately correct, it seems obvious that the Scrum model has found the most favor, but specific tools and approaches continue to be intermixed across the methods. The approach here is to not dwell on the brand names but to explore the management essence of these methods and try to see how best to use these in a general integrated model. It would be a futile religious-oriented argument to select one of these as being best. They essentially have a common core set of principles that spawned from the agile Manifesto or earlier prototyping ideas. One of the difficult aspects of this movement is the tendency of the designers to create special names and acronyms for various processes, tools, and objects. That statement is true of the whole agile community and it creates a somewhat cumbersome ability to develop a singular view. This essentially leaves a babble of new terms with internal bias within a core method. This internal fragmentation does not help broad acceptance. Nevertheless, we conclude this overview with a summary list of successful techniques noted from the use of iteration as a development technique:

1. Short work cycles
2. Small focused teams
3. The narrow scope of work
4. Active user involvement
5. Open and frequent communication
6. Focus on customer satisfaction more than schedule and budget
7. Use of a "sprint" to add urgency to the effort

One has a hard time refuting the items on this list as being positive attributes; however, it is important to keep in mind that this is not the universe of project management and that portion of the conflict remains active. Given the usage of Scrum as a favored delivery tool, more details related to this method are worthy of exploring.

Scrum and Kanban Methods

At this stage, Scrum is the most used dialect and Kanban charts are increasing in usage. Here is a brief definition of the two main dialects:

■ Scrum—the concept of a product backlog being packaged into fixed-time sprints represents a working definition.
■ Kanban—this method is based on early Japanese quality Just-in-Time manufacturing processes. The essential idea of Kanban is reflected in the visual chart of the same name showing work moving through the sprint.

From a methodology point of view, it is hard to distinguish significant differences between these two methods and there is now increasing cross-fertilization of their processes and tools

The formal Scrum life cycle is defined as:

■ Concept—What is the project trying to achieve and what are its "features"?
■ Inception—A team planning phase to determine work packages
■ Iteration—Sprint execution of defined work packages to produce planned features

- Testing—Evaluation of the sprint output to evaluate the status of sprint deliverables; results of this determine the status of features and this information is cycled back to the Inception stage for replanning the next sprint.
- Production—Finished products are moved to production
- Review—A review process to evaluate the status of deliverables and the overall progress of the project. Also, highlights, roadblocks, and lessons learned are shared.

The Scrum workflow is schematically shown in Figure 9.1.

The sprint-focused workflow outlined above is much like an assembly line with sprints being equivalent to the manufactured product. Packaged into the sprint is a planned collection of desired deliverable features. If a feature is not successfully created in a sprint, it is reviewed for recycling to another iteration. Alternatively, the user can say that the current level is adequate and that feature is considered complete.

There are other mechanical and process components defined in this methodology. For example, team and leadership roles are standardized. There is no title for a project manager but the *Scrum Master* leadership function is essentially the same. A *Product owner* is the business representative and is charged with making sure the deliverable meets the organization's needs. A *Scrum Master* is charged with clearing roadblocks and this is defined as "supporting progress." Note that the concept of authority is not visible here and the team role is viewed more like a self-managed structure. The view of individual work roles is one of the key differences from the traditional project team which typically contains a team leader and a project manager with more classic authority. Business representative roles may or may not be

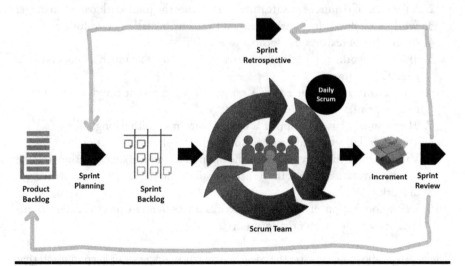

Figure 9.1   Scrum workflow

clearly defined in the traditional project structure but they are frequently actively involved.

As with so many features of a project management environment, Scrum contains a unique set of vocabulary and approaches to delivering output. These unique views are also noted in other agile dialects, even though the sprint execution process is similar in all. New users of this approach are faced with this set of unique features within their specific dialect and that opens up the increased potential for internal controversy in the future.

The term "Scrum" comes from the rugby analogy of team members huddled together to capture the ball. The growing popularity of this development approach suggests that the model characteristics fit something that has satisfied a broader audience. However, within all agile dialects, there is a focus on a "lean" approach to requirements definition before starting work. Deliverables are defined as features of the output rather than tractional fixed specifications. For example, a process feature might be to produce a named report (customer order entry). Specifics of this report would be left for the team and knowledgeable users to define further specifics and format. A machination of output features becomes the sprint team's work product. The typical schedule time for a sprint is two to four weeks, during which time the team is focused on that defined set of work. The size of the requirements planned for the sprint has been estimated to fit the planned time. During this cycle, the team evaluates a feature and proceeds to produce a suitable output. As with all of the agile dialects, there is a specified set of steps and procedures. The following steps represent the essence of the Scrum process:

1. The team consists of both technical and knowledgeable users who are interested in the output.
2. A backlog of requirements (features) represents the total workload to manage.
3. Work is produced in a Sprint-defined time box (usually two to four weeks).
4. A sized list of features is allocated to a sprint.
5. The team works on specified features and produces as much as possible in a fixed time frame.
6. Any feature not produced by a sprint will be replanned and moved into a future sprint.
7. The team is focused on sprint activities, with no multitasking.
8. There is a daily, brief meeting to communicate short-term actions.
9. Various status tools are used, but the main one is a Kanban chart showing the short-term status of work in the active sprint—no dates, just the flowthrough of work.
10. Upon completion of the sprint, there is an evaluation process called retrospective which reviews the sprint cycle.

The ten steps above summarize the basic Scrum work process, but keep in mind that the overriding question that is relevant for this discussion is "Why are these types of methods being viewed positively?" That is a much tougher item to accurately summarize and the answer is based on one's philosophy of how best to achieve a delivery goal. The following five characteristics seem to offer the most reasonable explanation for the positive view:

1. The team is given more *flexibility* and *delegation* in planning and delivering work.
2. There is pressure to produce output within the short sprint cycle (*faster delivery*).
3. Multitasking is decreased as the work does not have slack time estimated (*productivity*).
4. The sprint team represents a *coherent work group* that can be very productive with common goals and delegation of tasks.
5. The daily standup status process should be done in all projects (*active communication*).

Whether this is the global answer to project delivery or not, there is clear evidence that many of the work process elements described here are related to improved customer satisfaction and shorter delivery cycles. As with all the models described, the goal is to not get excessively lost in low-level details but to focus more on understanding the management logic of the model.

Modified Scrum

Based on the positive deliver characteristics outlined, there is a strong indication that the use of iteration techniques will be increasingly expanded over time into the predictive environment even though much of the traditional industry has claimed that this is not appropriate or doable. A lot of things are doable whether they are appropriate or not. The authors are proposing that an iterative approach can be utilized in the predictive arena with a slightly revised process. This method is titled Modified Scrum. It is envisioned as a companion work option to the traditional waterfall task approach, and it can be embedded into a tractional plan as a dual work method. For this option to fit into an integrated delivery model, this has to be able to incrementally snap in when the conditions fit.

Modified Scrum sprint work is to be executed on graduated requirements. The recommended approach for this is to use a multi-level requirements definition feature titled *MoSCoW* (i.e., Must, Should, Want). In this case, the requirements would have multiple levels of acceptance, which allows the sprint to have flexibility in delivery. The required level of a sprint's output would be set by the MoSCoW level definition and based on the task characteristics. The sprint is then scheduled and executed just like classic Scrum. Sprint scheduling has one other required process modification. Since this sprint is producing a tangible product that is not decomposable, it has to complete at least the minimum defined MoSCoW requirement level. If that is not done, within the scheduled time, the sprint has to stay active and overrun until that level is achieved. The use of these graded requirements converts the process into a feature-like view and makes this a feasible work option. The following list summarizes the steps described:

1. Use the MoSCoW graded requirement approach for the target work units.
2. Define the minimum acceptable completion level for the units.
3. Estimate the time needed to produce the deliverables target.
4. Package the work units into sized sprints.
5. Schedule the sprints to mesh with the traditional task plan.

The Modified sprints will be internally managed using various agile tools such as Kanban charts. And the overall view of this will be shown in the traditional plan structure. Even though the sprints are embedded in the tractional plan view, they represent a dual work management flow and are executed using standard sprint techniques. From a management viewpoint, the most visible difference this approach brings is a modified view of work schedules, although more differences will be found as other processes are modified in the integrated model. Since the Modified Scrum process is considered a published model modification, a more detailed example of this process is shown below.

Modified Scrum in a Traditional Plan

This section will demonstrate a simple approach to embedding iterative work units into a traditional project predictive structure. Figure 9.2 shows a traditional WBS with two work units, 1.3.2 and 1.4.1, defined for execution as Modified Scrum sprints.

Using the WBS data in Figure 9.2, a traditional first-cut project plan can be produced as described in Chapter 8, then modified to fit the new work option.

Figure 9.3 shows the modified dual work option project plan. Two manually scheduled sprints are shown for WBS IDs 1.3.2 and 1.4.1 and they are embedded in the plan with hatched bars to show that they are executed differently. Also, based on their roles in the overall plan, it is hypothetically decided that an overrun at 1.3.2, Design 2, would be disruptive for the follow-on Execution work. So, an

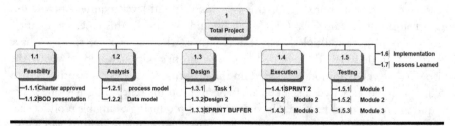

Figure 9.2 WBS with embedded sprint work units

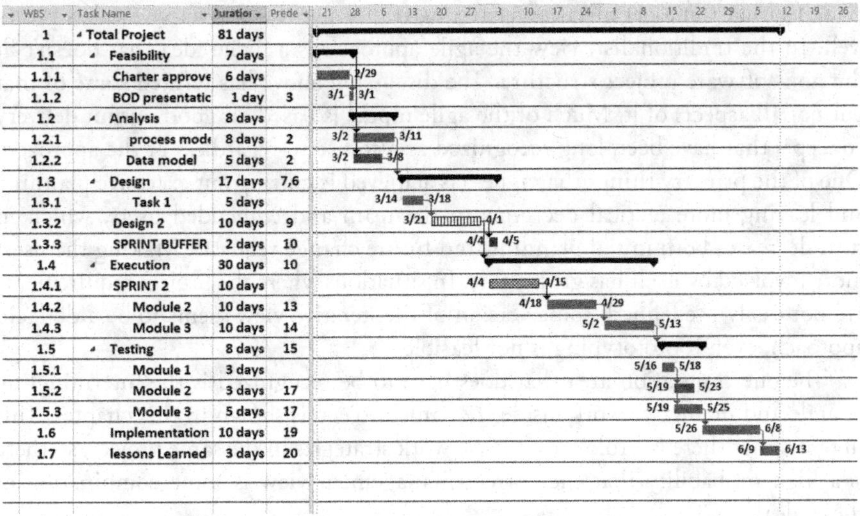

WBS	Task Name	Duration	Prede
1	⊿ Total Project	81 days	
1.1	⊿ Feasibility	7 days	
1.1.1	Charter approve	6 days	
1.1.2	BOD presentatic	1 day	3
1.2	⊿ Analysis	8 days	
1.2.1	process mode	8 days	2
1.2.2	Data model	5 days	2
1.3	⊿ Design	17 days	7,6
1.3.1	Task 1	5 days	
1.3.2	Design 2	10 days	9
1.3.3	SPRINT BUFFER	2 days	10
1.4	⊿ Execution	30 days	10
1.4.1	SPRINT 2	10 days	
1.4.2	Module 2	10 days	13
1.4.3	Module 3	10 days	14
1.5	⊿ Testing	8 days	15
1.5.1	Module 1	3 days	
1.5.2	Module 2	3 days	17
1.5.3	Module 3	5 days	17
1.6	Implementation	10 days	19
1.7	lessons Learned	3 days	20

Figure 9.3 Project plan with embedded sprint units

overrun buffer is inserted after the sprint at ID 1.3.3 to protect that phase of the project. More discussion regarding the role of schedule buffering is upcoming in Chapter 10 and beyond. For now, just look at this as a schedule protection strategy. It is also decided that no buffer is deemed necessary after the 1.4.1 sprint (management decision).

This example shows that the two work methods can coexist in the same plan and still be managed differently. Each sprint would follow the agile process mechanics and use the graded MoSCoW requirements process that allows a single sprint cycle approach. Chapter 10 will describe further the use of buffers as a schedule protection idea.

At this stage, we are restating an earlier design conclusion outlined in Chapter 3. Based on the variances found in the target goal, not all work methods are necessarily the same across the project. So associated work methods need to be flexible in matching those characteristics. In this example, some of the work units did not have sufficient requirements definition to match the tractional model, and the Modified Scrum process is employed. We will define this matching idea as *Work Management*. The ability to mix and match multiple work types such as the traditional and iterative views within the same project structure is a key part of the integrated model design goal. There are still more work options to be examined but this example gives a good perspective for showing the dual workflow process.

Is Iteration Productivity a Myth?

There is sufficient evidence at this point to conclude that the concept of iteration productivity is not a myth. The real question is why it seems to work so well while the traditional approach continues to struggle. On the counter-side, there is evidence that some organizations have learned to use the predictive models pretty well. In the traditionalist, view the agile approach is a niche idea that doesn't fit the non-software project structure. The discussion above has refuted some of that but not all aspects of it. Much of the agile model is based on good output delivery concepts that have been long recognized as also useful regardless of the approach. One of the primary things that agile has achieved is cutting out excessive planning and leaving more tactical decisions to the team and embedded users, which in turn decreases both initial planning and future change requests. Making the users more involved by itself has great value. In situations where the project requirements are not easily predefined, some version of the iterative view seems to be a de facto approach, even if prototyping is not feasible.

The one translation area that does have to be recognized is the intermingling of agile and predictive work inside the same project as shown in this chapter. This suggests that there is a role for multiple work strategies in the same project, so there is a high probability that the correct management view is some combination of these ideas.

Conclusion

Nothing has changed the project management landscape over the past 20 years more than the concepts surrounding the term "agile." This chapter has summarized the delivery attributes of this model as well as gaps in the overall life cycle coverable. Techniques related to agile, Scrum, and Kanban have shown how this delivery method has evolved over a relatively short cycle. Survey data and user case studies have highlighted significant improvements in customer satisfaction and project cycle times when using this method. Conflict within the industry remains concerning the acceptance of these methods for traditional predictive work but there is evidence that this will be challenged over time.

The current view of producing project output via iterative techniques is shown to be more than the original simplistic software prototyping view and this concept continues to morph into broader development areas as the internal processes mature. The use of fixed-time sprints to execute work is being increasingly popularized and it has shown positive team productivity attributes that need to be understood by the contemporary project manager. The traditional model suffers from a static view of work that does not encourage action, while the iterative form provides more of a focused action culture.

The iterative development process works well in software projects based on the ability of that deliverable to be decomposed into chunks of usable output. This means that the user can examine actual partially complete versions and from that better decide how to modify that for the next sprint cycle. This process provides faster output delivery.

After the initial publication of agile principles in 2001, acceptance of the iterative model became the de facto approach for software development and this usage continues to expand into larger and different project types. Industry bias still views iteration production as a niche technology that is not suitable for a tangible product that can't be delivered by sprint mechanics. The Modified Scrum dialect model is shown as a potential candidate for using iterative techniques in a predictive work environment when the requirements cannot be accurately predefined.

References

Agile Manifesto Authors. 2001. www.agilemanifesto.org (Accessed July 15, 2022).
Distant Lab. 2022. Agile Methodology. https://distantjob.com/blog/agile-software-development-life-cycle/ (Accessed November 1, 2022).

Further Reading

For the reader interested in more iteration model detail, the following are useful resources:

- Agile Alliance (www.agilealliance.org)
- Hoegl, M. and H.G. Gemuenden, "Teamwork Quality and the Success of Innovative Projects: A Theoretical Concept and Empirical Evidence" in *Organization Science*, Vol. 12, No. 4 (Jul.–Aug 2001), pp 435–449.
- Scrum Guides: (www.scrumguides.org)
- Spiral model: https://www.sciencedirect.com/topics/computer-science/spiral-model

Chapter 10

The Critical Chain Model

Note: Material from this section is adapted from Richardson, G.L. and Jackson B.M. 2019. Project Management Theory and Practice, 3rd ed., CRC Press.

Reader Note: This chapter's description of the Critical Chain (CC) model focuses only on the core processes that have the most universal value to the integrated model goal. The breadth of buffering outlined in the full CC model is beyond the scope of this text but has potential value to the project manager and is worth understanding from a more global view. In the author's view, CC theory has significant management value well beyond what the industry has understood. It contains a management discipline that clearly illustrates how disciplined resource management can significantly improve schedule performance.

Introduction

The role that the Critical Chain (CC) model has in the overall text scheme is most associated with time compression much like the traditional model idea of crashing a schedule. CC theory is one of the more interesting contemporary project planning concepts. This model is based on the application of Dr. Eliyahu Goldratt's *Theory of Constraints (TOC)* (Goldratt, 1997). CC's approach to project planning and execution requires project managers to abandon traditional estimation and project control practices. Management of the CC elements is handled by the use of resource alerts and buffer management for defined work chains. Implementing these concepts will require a cultural change throughout the organization as fixed task schedules are thrown out of the plan, Critical Chain Management (CCM) looks at projects in a new light, by changing the way projects are estimated, scheduled, executed, and controlled. In an ever-increasingly intensive environment, the management of projects, particularly product development efforts, is

DOI: 10.1201/9781003431091-12

increasingly one of the factors that can produce a sustained competitive advantage. Firms that can bring products to market faster than their competitors can extract higher initial market share and margins. The underlying theme of this model is to complete prioritized projects faster and to make more efficient use of critical resources.

The focus of the CC methodology deals with four organizational processes and cultural problem areas:

■ Conservative task estimating—padding tasks to preserve variations
■ Worker procrastination syndrome
■ Multitasking—too many work tasks in play at the same time.
■ Next-step resources are poised and ready to start as soon as the previous task is completed.

We will show how each of these human behaviors affects project outcomes.

Basic Setup Steps for the CC Model

Kendall et al. (2005) summarized the following list of concepts and mechanics used to manage the CCM process regarding people and tasks:

1. Task estimates do not have padding. They are planned at some probabilistic completion level, such as using estimates for which there is a 50% probability that the task can be completed at the defined time.
2. Project team members are dedicated to being available for their assigned tasks and work on that task until it is completed. Periodic status reports are required to indicate the time remaining for each active task. Every effort is taken to eliminate delays and work procrastination.
3. Multitasking is eliminated by assigning workers to tasks in priority order and completing that task before moving on to a new task. Industry experience suggests that multitasking creates inefficiencies amounting to as much as 40%.
4. Managing tasks by the due date are not followed. Workers and tasks are not measured based on scheduled completions. The management approach is to pass on the task to the next activity as quickly as possible. *This is a track meet metaphor.*
5. By taking resource dependency and logical dependency into account, the longest sequence of dependent tasks can be seen more clearly. This longest task chain sequence may cross logical paths in the plan network. So, the management view is to deal with resource status across the chain and not just with critical path schedule management.

6. Buffers are a key mechanism to manage desired schedules. These will be defined to protect various aspects of the project schedule. Project status will be monitored by evaluating the status of buffers.
7. Every effort is made to create this culture in the project team.

Note how the processes outlined above affect three target areas of task padding, procrastination, and multitasking. Each of these points is somewhat subtle as to their impact on project outcomes. From this list, we can examine how each of these practices affects the project schedule.

Resource Discipline

Managing the timely availability of resources is a known general best practice, but the traditional discipline for doing this does not equal the rigor specified in the CC model. Not having a resource available when a task is ready clearly expands the schedule beyond what it would be without the gap. To accomplish this level of readiness, a dynamic resource alert process is needed. Not only is the CC goal defined to have the resource ready when the predecessor task is finished, but it also wants it to be ready in advance and have any prep work finished before the start of work. We will show here that the CC-defined process brings a new level of resource management discipline to the project and this approach should be considered a best practice for any model.

Task Padding

This issue has been mentioned repeatedly as one of the evils of project scheduling and control. The CC model deals with this issue as a core process and offers clear evidence as to why it is viewed as described. The issue of task padding to avoid overruns on the surface would seem to represent valid logic but there are hidden repercussions that make this a poor practice. One technical root cause of task over-run comes from resource skill differences and a host of other such factors. These are accepted as reality factors; however, a major segment of such overrun occurs because of resource behavioral factors that seem to have been largely ignored in practice. The logic behind the task padding culture was described in Chapter 8 so we won't repeat that here.

Procrastination

Related to the issues above, human procrastination triggers the overrun process because the worker either is busy and feels like there are other tasks more impor-tant, or there is a perception of such. The net result here is the padded task is still overrun despite the initial padding. This is a psychological trait that causes more projects to overrun than most understand. Even though this approach to task management is being described as part of the CC model, it is a project culture problem. Based on the recognition of these human traits, the CC model requires that ask estimating is even more rigorous than using a raw estimate. In this model, the estimates used are called 50/50, meaning that the time estimated indicates only a 50% chance of completion. This changes the dynamics of the schedule. Overruns are now the norm so the traditional concept of project tracking is lost. Beyond that, there is a need for action and a major stimulus for speed that simply does not exist in the traditional project culture. This represents the power of the CC view.

Multitasking

Various studies have shown that when a worker attempts to do multiple tasks at one time there is greater inefficiency than if each task were pursued in priority order with no interruptions. CC pursues this idea just like a track relay race. If a track runner is trying to make the best time, they do not stop in the middle of the race and do something else. What this means in practice is that there will have to be a defined priority for tasks. In the CC view, it is based on the project plan and a formal assignment of the resource to that task. One can see that several cultural changes would have to be made in the typical project to implement this type of execution environment.

Internal Factors

The CC method focuses on managing project network task chains with restricted task duration, buffers to protect overruns, and disciplined resource management across the project. This discussion includes a general explanation outlining how CC principles are applied to project management to construct the resulting plan and underlying process. Also, the implementation complexity of the model and some of the main challenges faced will be described.

Work Chains and Buffers

The architecture of the CC model is based on optimizing the execution of sequential task work chains. From a conceptual standpoint, this management technique has a work flavor reminiscent of agile/Scrum sprints. Another application of this concept is found in a task group called a Work Chain whose working definition is:

> A collection of task groups that have been chosen for expedited completion. This task group will be rigorously pursued from a schedule and resource point of view for minimal time completion. Within this grouping of defined tasks, resources are singularly focused only on that

target using CC management logic. This strategy can also be applied to the entire project.

In the case of a predictive project plan, the task chain could be shown schematically as in Figure 10.1, which shows a classic predictive CC structure, with a critical chain and a feeding chain. This view fits the standard CC model for time compression; however, an alternative view of this could be to segment this into a subset such as tasks Y and Z, and only execute that portion with CC principles. A third choice is found in any segment of the project plan that needs this management approach.

Assume that the boxes in Figure 10.1 represent tasks with CC time estimates and resource requirements. In this example, the task list represents more than a single resource skill as represented by the various icons. Think of such a task grouping as a focal point or a critical "chunk" of work. Remember that the Goldratt theory recommends that task estimates be made at a 50/50 level so there is no slack time in the work chain. This example is taking some liberty with Goldratt's buffer logic and will only define a single chain buffer rather than the more sophisticated other buffer types that tend to complicate the process. A significant portion of this model's advantage is achieved by the estimating and work management aspects, so this is a manageable segment of the defined method. More can be achieved if one wishes to deal with the other buffers outlined by the theory. If we were method purists, there would be another buffer at the end of task Q to protect that sub-chain group from the main chain. The CC model is very liberal with the use and types of

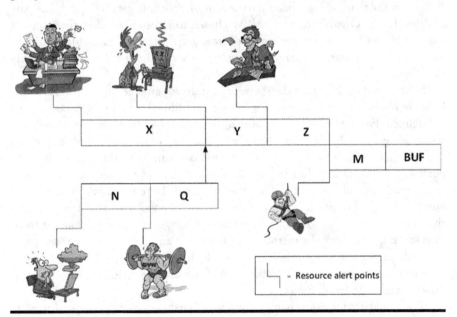

Figure 10.1 Sample work chain. Source: Richardson, G.L. et al. Used with permission

buffers. Also, if a single resource type is used for all tasks, then another buffer might be required to protect the availability of that resource.

The use of work chains and buffers as described here plays havoc with the traditional plan versus actual control schemes. In the traditional plan, one can view this collection of tasks as having a schedule for completion. Contrarily, in the CC model, the schedule is to complete the task grouping as quickly as possible and hopefully before the buffer is consumed. This is a different mindset of control. In the traditional model, the project plan can be compared to a bus schedule, while a CC view is more like a track relay race with everyone running as fast as possible. CC theorists claim there is no motivation to move faster in the padded completion times environment. If a task is scheduled to start on Wednesday the only issue is to try not to overrun that time—i.e., stand on the corner and wait for the 8:15 bus to arrive. Another subtlety of the CC work process is the increased motivation to finish a task. In this modified essentially unscheduled view, there is no schedule for the bus to arrive. It is driving the route as fast as possible and the riders need to be waiting for pickup. This alteration in project control logic will challenge the traditional project manager and other organizational levels.

Resource Allocation

The second aspect of chain logic involves how resources are managed. Once again, we have to say that having a fixed start and finish schedule for each task is not part of this scheme. Organizations that have chosen to rigorously follow the Goldratt model have had to change the way resources are made available. This requires a more dynamic resource management system as we will illustrate using the schematic model.

Note in Figure 10.1 that the resources are shown attached to each task. What is not so obvious from the schematic is that a scheduled time for each task is now not defined. For the chain logic to work as defined, it will be necessary for assigned resources to be available when the task is ready to commence. Think of a chain as a relay race with a baton used to pass from one runner to the next in line. We don't know exactly how fast the previous runner is, but the next resource needs to be standing ready as soon as the baton is passed. Here is the key philosophical point. Is the goal to finish this project as quickly as possible or not? If it is, this is the essence of a task estimating and resource management process that has to be in place. Implementing the methods described will create significant challenges to the traditional project culture. Whether work chains are designed into the project plan structure or not, the timely allocation of resources to assigned tasks is a high-priority challenge for all projects.

The actual resource mechanics are not as harsh as described here. There are many things that a resource can be working on. Some of these could be simple prep work for the upcoming tasks but other items fall into the category of lower non-time-sensitive priority. The key management action is to look at the chain task as the highest priority. Remove all obstacles and time consumers while the chain task is active. No multitasking. When the relay runner is executing their leg of the race, they do not stop for anything outside of the specific task.

Earlier, the statement was made that project performance improvement required more understanding of the critical roadblocks and the use of new management processes. The CC model offers some of the best examples to illustrate how work management can affect outcomes. Defining and managing work chains is certainly just one clear example of this.

Resource Alerts

Once the CC structure is established, the problem turns to resource allocation and related conflicts. In Figure 10.2 a buffer is inserted at the end of task N8. This is done to ensure that this sub-chain does not overrun and affect the critical path chain cycle time.

As work progresses, the individual task forecast and buffer status become the primary status metrics for chain completion. As an example, when a predecessor chain task resource reports planned completion the next stage resources move into position with five days remaining (Goldratt recommended). As the predecessor task is completed the next task is started immediately (think of a track race baton passing here). In the relay race metaphor, this is analogous to getting ready to accept the baton. This mechanism represents a dynamic countdown for the successor. If the predecessor reported two days later that there were still five days remaining, this

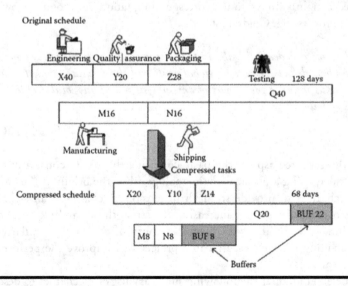

Figure 10.2 Comparing traditional vs CC cycle times. Source: Richardson, G.L. et al. Used with permission

information would be passed on to the successor, and a review of the start-up steps for the next activity would be performed. Note in this process that the focus of all activities is on having resources ready and prepared to execute the next task and then move as quickly as they can to completion. Tactically, the focus is on ensuring that resources are in place as required.

The final descriptive step of CCM involves how the schedule is managed. In summary, project status is tracked primarily through the status of its buffers. This discussion did not describe the full set of buffers defined in the model, but the concept of protecting the work chain completion is the key design theme and buffer consumption status offers clues as to how the chains are doing. Project status reports would focus on these. More sophistication can be added by calculating buffer "burn rate" so that if one is decreasing faster than task completion it would trigger to review of that chain to see what needs to be done. The buffer management process highlights potential problems much earlier than they would ordinarily be discovered using typical project management techniques.

CC Design Concepts

The key design focus for CC is increasing the work chain completion speed but speed must be achieved without compromising other aspects of the project deliverables such as service, features, or flexibility. As the demand for shorter project cycle times and more deliverable flexibility grows, so does the frustration level of both PMs and their team members. According to the model, a project's failure to deliver as planned is tightly linked to the corresponding failure to recognize how the task chains are progressing. Goldratt stated:

> *Before we can deal with the improvement of any section in a system, we must first define the system's global goal; and the measurements that will enable us to judge the impact of any subsystem and any local decision, on this global goal.*

(Leach, 2005)

Beyond the resource aspect, the basic concept behind CC constraint theory is best described using a physical chain analogy (thus the origin of the name). The goal of a chain is to provide strength in tension. A chain's weakest link determines its overall strength, so increasing the strength of any link other than the weakest link will not affect the overall strength of the chain. Similarly, consistently managing the weakest link in a project will improve the performance of the overall project.

Herroelen et al. offer the following summary list of mechanics to describe the fundamental CCM process (Herroelen et al., 2002):

1. 50% probability activity duration estimates
2. Task chains defined with no due dates
3. No fixed project milestones and no resource multitasking
4. Minimize work in progress (WIP)
5. Define the project in terms of a network structure as in the traditional view\
6. Identify the critical activity chain
7. Insert appropriate buffers to protect the defined task chain
8. Keep the baseline schedule and CC fixed during project execution
9. Determine unbuffered schedules and track completion schedules
10. Use buffer status as a proactive warning mechanism during the execution

This view differs from the classic critical path network definition in the waterfall model. The waterfall view only reflects the fact that a defined task exists and is linked to other tasks, while CC recognizes how related resources are assigned to the tasks across the project. The CC view is conceptually similar to traditional capacity management, but with a more dynamic flavor. In the CC view, resources are managed dynamically based on the current status which better supports faster project completions. By doing this, the task chain completes faster and generally decreases overall project cost as a by-product. The underlying subtlety of this statement is there is no defined fixed completion time and the resources are managed to be available as soon as needed. Think of this as a relay track event with the baton representing the completion time for the next task. Everyone in the chain is poised to move each task as fast as possible with no schedule gaps.

Role of Buffers

Figure 10.2 illustrates the time compression resulting from the CC method. Note the top original project plan with padding shows a schedule of 128 days. The bottom version shows the tasks estimated as described using CC logic. Two buffers are inserted—one to protect the subchain overrun and the second to protect the project completion. In this example, the schedule is essentially cut in half. A more realistic estimate is to cut the overall project schedule by 20%.

Looking at Projects in the Real World

As a professional project manager, it is hard to look at so many poor management examples in the real world that exhibit a poor focus on goal achievement. One common example of this is the infamous customer service system where one waits in line for 45 minutes to eventually obtain "service." Many of these real-world examples show that a project likely did create this result, but the true goal was lost in translation. One of the personal problems of being exposed to project management principles is that inept use of these management principles stands out when one deals with a failed example. The question emerges, "Was this management process defined to simply say we had something?"

Road construction projects are the one global example that all drivers experience firsthand. This is a classic predictive project model with well-defined scope. If there is a project type that fits the traditional model, this is it. So, is your impression that they are typically properly managed? Our thesis is no! Are they finished on schedule? Are the driver stakeholders considered in the way work is performed? Does it appear that resources are applied in any fashion similar to what is described here? You can make up your conclusion to these questions but there is an argument that these projects have management gaps. This environment was chosen to offer an example that may help one better understand the management of work chains and resources.

Experience says that the task logic for many road projects is to lay out a long line of work (a work chain) and then spend a considerable period working on that one long chain. A lot of orange cones and slow traffic is locked around the area for safety reasons possibly. The drivers would like to see less congestion and quicker completion. So, how can chain and resource management logic support that goal? The first speed-up goal would be to divide the total effort into smaller work chunks and focus on those using chain logic. It may well be more efficient to pave three miles of a road as one long work chain, but it is certainly not optimum for the driver's side of the equation. In many cases, such projects look like they are designed for the contractor and not the customer/driver. Work chain logic using essentially agile and CC work principles would be of benefit to all parties.

A more interesting example was found recently in a local suburban construction project. This project is already two years late for undefined reasons. Traffic is queued daily at one key intersection with heavy use of a turn lane—70 cars in one lane, static at a traffic light. At this observation point, no work was underway for a one-mile strip adjacent to this intersection. If we were to look at this in a traditional plan, there would be work units for drainage, surface, reinforcing bars, concrete pouring, etc. Each of these might be laid out as sequential linear tasks for resource efficiency reasons. Each of these long-overlaid work segments might take multiple months to complete while at the same time leaving the total road unusable and the intersection clogged each day. During this time, the same 70 cars would get to queue each day at the same traffic light awaiting the total road to be finished. This is an example of a work "pinch point." If 300 feet of the road would be completed at the traffic light turn lane, the daily queue would decline significantly. Also, if this 300-foot section of work was converted into a high-priority CC-type work chain as described here, there would be a very positive outcome. In observing this same point some six months later, it remains unfinished with work going on in a non-critical segmetn).

An opinion point regarding this scenario is to suspect that the indicated 300-foot segment is likely not on the traditional plan critical path. Recognizing pinch points within the project plan could improve deliverable success with little change in required resources. These are segments in the project that can be separated to deliver early to improve customer satisfaction. The key point of this example is to suggest that the best goal of any project is not to just focus on a defined critical path but to look for segments to improve user satisfaction, maybe even at the cost of a slight schedule or budget increase. A project goal is not always just trying to finish the critical path but also should look for reasonable ways to improve customer value even before the project is completed. The existence of pinch points can be found in many projects.

We will try not to beat up the construction industry with these marginal management examples but they make such good visible ones that it is hard to not use them. One final comment on the above project. While it is interesting that the current project locks up maybe 70 cars at the traffic light because of an unnecessary lane closure, the cross street is the main thoroughfare with even more traffic. Thousands of cars go in that direction and they also have to wait at the same light for similar reasons. It is interesting to ponder this situation while waiting on the new street to be finished (all these months). One thought occurs from this is which road should be given priority work and how should the overall intersection be managed. If the main road was improved first, much of the alternate street traffic would go around. The real source of the traffic congestion is the lack of an overpass for the nearby railroad crossing and the referenced side street. One could argue that this entire current repair project is a band-aid that does not address the root problem. So, the real question here is what should the project scope have been? This example illustrates the role of portfolio management and we need to delay more discussion regarding this topic area is deferred until Chapter 12. As indicated earlier, project management issues occur all around us and they affect not only our traffic, and decision but also many competitive issues in organizations.

CC Organizational Challenges

This chapter has presented a project model that has great potential, although culturally difficult to implement. CC brings with it new processes and concepts that do not fit well with traditional status views. The following list represents eight of the most significant challenges that organizations will find in implementing CC concepts:

1. *High-level management support:* As with all significant organizational changes the support of senior management is paramount. Their fixed status sheets will be greatly affected.
2. *Cultural change in managing teams and projects:* CCM changes how project activity is pursued. The entire organization must understand and work with the new paradigm.
3. *Status reporting methods:* Traditional status reports will have to be replaced and all stakeholders will need to be educated on the new approaches. There

will likely be the need to compromise on methods related to completion reporting. There are many related status-oriented changes embedded in CC processes. It is important to remember that effective communication with the stakeholder community is also a prime goal for the PM. This change will not be a transparent one.

4. *Translate estimating techniques to 50% probability:* Taking away time padding will be a major cultural problem because of the stigma of time overrun in the traditional view.

5. *Task overruns are now the norm:* Traditional status reporting looked unfavorably at time overruns. In the CCM model, they are expected. Management and other stakeholders will have to understand this new phenomenon.

6. *Team evaluation:* In a relay race, the team wins and that is the way CCM must work.

7. *Resource allocation and project priorities:* Resource alerts and formal project prioritizations are required to manage the workflow process. Both of these issues require more discipline than exists in the typical organization.

8. *Multitasking avoided:* This implies that once a resource is moved to a task, it will work on that task until it is completed. No jumping around to other tasks.

As indicated, CC concepts can be viewed as radical. This approach clearly will make the project environment high-speed beyond the norm. Critical Path has a lesser meaning now as everything is critical in the chain—no slack times and resources are viewed more like firemen ready to respond to the bell. Offering this method to an organization with the promise of reduced cycle times will be an operational challenge because it attacks many of the cultural morays. As with an agile-type introduction, some evidence of success would seem to be required before acceptance would follow.

Conclusion

The CC model represents a viable option to improve project deliverable times by as much as 25%. If that were to be verified with a few test projects, doesn't it make sense that this might become a very common approach to predictive-type projects? In any case, this approach offers interesting insights into why projects are perceived as being too slow. Even though the processes described here for the CC model seem unusual, they are more of a logical extension of traditional project management practices than it first appears. Be aware that trying to implement the full range of the CC model is more of a challenge than described here, but the real value found here is understanding the time compression essence of this model and the value that even portions of it can bring. One should look at the basic concepts summarized here to see how they might be implemented by first testing the approach on a pilot project. It should be clear with this review that there are excellent project management ideas embedded in this model regarding methods for improving project throughput. Certainly, the use of buffering and restricting time estimates could be implemented in some fashion in all projects.

Many organizations today are searching for better ways to achieve breakthroughs in project development cycle times to stay competitive. The goal is to push more projects through the organization per unit of resource allocation. This goal must also often be achieved without increasing the number of people allocated to projects or having the option of hiring additional people. The availability of skilled resources will always be a project constraint in both good and poor economic times. In healthy periods, the aggressive requirement outstrips demand, and in tough economic times, executives are reluctant to hire even though the demand for new projects remains.

Out of all the more formal project management schemes proposed in this text, the CC design execution logic is probably the best thought out from a conceptual point of view. The uniqueness of this model hits at the heart of why projects take too long to execute. No matter what project management process is chosen, so long as padded estimates, procrastination, and multitasking remain the cultural norm, projects will continue to overrun as they do today. The logic underlying the CC model concept is so compelling that the modern PM must understand both the power and operational complexity of this model. These basic concepts must be added to whatever future task management approach is defined.

References

Goldratt, E. M. 1997. *Critical Chain*. Great Barrington: The North River Press.

Herroelen, W., R. Leus and E. Demeulemeester. 2002. Critical Chain Project Scheduling— Do Not Oversimplify. *Project Management Journal*, 33(4), 48–60.

Kendall, I., G. Pitagorsky and D. Hulett. 2005. Integrating Critical Chain and the PMBOK® Guide. http://logmgt.nkmu.edu.tw/news/articles/criticalChain-PMBOK. pdf (Accessed April 12, 2008).

Leach, L. P. 2005. *Critical Chain Project Management*. 2nd ed. Norwood, MA: Artech House. Pp. 23, 53.

Chapter 11

Organizational Support Architecture

Flower Bed Metaphor

Most projects are executed under the purview of a host organization. There are many ways in which the organization's operational role can be described but the one that seems to be most memorable is to look at this relationship as one of a flower bed and a seed. In this metaphor, the project is the seed that hopes to grow into a beautiful organizational flower. The organization is the bedding environment that will support the seed's needs as it grows to fruition. The flower bed (organization) provides water, fertilizer, weed control, and sunshine to the project. Absent a supporting mature flower bed, the seed (project team) would have to create the entire flower bed, which would require extra resources that do not directly contribute to executing the deliverables. A model organization would have a complete support environment for its projects. So, what does this strange metaphor mean in reality? In mechanical terms, it often means that a work facility is allocated to the project with various generic support items. In many cases, the project team resources are also drawn from the host organization and this is the epitome of support. Beyond the raw resource support, many other process systems aid the project. Failure to have these leaves the team to invent more crude standalone versions for their needs. For example, the organization's material inventory system is a valuable asset for procurement needs, including purchasing and storage. Assuming the team resources are allocated from the host organization there is the need to handle all aspects related to that resource, including payroll, retirement, insurance, and other HR functions. Information technology architecture is increasingly complex and provides high value to the team (i.e., hardware, software, Internet,

DOI: 10.1201/9781003431091-13

telecommunications networks, data storage, etc.). The organizational accounting and finance system is also vital to the cost management side of the project. Various project-related systems can be shared across all projects, such as change control, email, specialized software, and standard templates, which are all major assets to the team. The major point of this example is to understand that a project team most often does not start with a "barren field" with no support from the organization. As an organization increases its use of project initiatives to achieve improvement, this support role is more important than many seem to understand.

The concept of organizational support for projects is not a new idea but unfortunately seems to be viewed as mostly an academic subject rather than an active goal for organizations.

Organizational Maturity

One of the terms used to describe this concept is organizational maturity and several consulting-type organizations describe models for this. The theme of these models is focused on evolving through stages of operational effectiveness (i.e., maturity). There are important ideas described in such models that make it worthwhile reading to understand the multi-stage evolutionary process. These stages describe topics such as:

- Standardized processes
- Project data stores
- Portfolio analysis of project proposals
- And much more.

The concept of maturity relates to functionality in its environment. In the case of project maturity, the definition relates to those capabilities needed by the project. The term used to define this is *organizational maturity.*

There is now recognition that projects inherit much of their operational and resource support from the host enterprise and the maturity level of the organization in turn affects project performance in subtle ways. Secondly, organizations pursue a wide array of projects at any one time, all of which are competing for the same enterprise resource pool. Both of these situations impact management actions necessary to produce successful outcomes. One implication of this view is that a project can be more effective if the host organizational processes support its needs. If not, the project will be negatively impacted as it must use internal resources and time to deal with these and other similar issues. All of these support functions fall under this umbrella. The basic theory underlying this idea is that the higher the maturity level, the more capability the organization has to achieve its goals. "In the project case, this means better support for its environment and statistically higher outcome success rates" (Richardson and Jackson, 2019, p.499).

Over the past several years, various maturity models have been defined to describe specific ideas related to this term. One of the most recognized maturity models was originally published in 1986 and titled the Capability Maturity Model (CMM) from the Software Engineering Institute (SEI) (Masters and Bothwell, 1995). A second well-recognized maturity specification published in 2003 is PMI's Organizational Project Management Maturity Model (OPM3). (Farrokh and Mansur, 2013). In addition, there are several similar definitional models produced by various technical organizations. All of these descriptors are worth reviewing as background material. The CMM description is judged to be the most used and the most robust of the set. As is typical, it has a five-level scale that can be used to measure an organization. On the negative side, it is focused primarily on software development, while the OPM3 description is more general. Several other maturity offerings in this vein focus on other organizational components or processes.

The OPM3 model was intended to be a definition of an evaluation technique for achieving improved maturity, complete with certification. It contains definitions of best practices and measures of maturity so that organizations could see the areas that needed improvement.

What is interesting regarding the topic of organizational maturity is how little visible evidence there is now for pursuing this. The concept of portfolio analysis for the selection of projects is one of the few focus areas that have remained active but much of the maturity definition topic list seems less visible. It would be conjecture to explain this lack of visible concern but the topic area is legitimate. All organizations have some degree of maturity and this factor has been statistically measured as having an impact on project performance (Ibbs, 2002). Ignoring this aspect of project delivery leaves a gap in our search for techniques to improve delivery success. An understanding of this concept serves an important role in our integrated model approach.

Project Support Architecture

If one looks at an organization that is using the project to produce desired outcomes, it begins to make sense that integrating its activity into the fiber of the organization is key. In the case of the integrated model, there are three major organizational touch points to the project. These are senior management's role in selecting projects (discussed in Chapter 12), organizational resources needed to staff projects, and the general project support architecture. In quality management terms the project is an agent of the organization and likewise, the organization is an agent for the project. These are two mutually supporting roles. Performed efficiently, there is synergy for both. In the ideal case, there should be a smooth linkage between the three elements described. Resources and organizational processes should be readily available to provide needed project support.

One way of looking at the organizational maturity question for project support is to evaluate common processes that a project needs and supply them with minimal bureaucracy. If one is attempting to create this environment, it is important to have a visible and formal approach to that activity. Ironically, the process of creating this organizational environment will become a multi-phase project plan in its own right. All required support processes are viewed the same way that the development decisions are. The design goal for these is to produce such images as a Lego-like block that can be simply "snapped" into to obtain the needed service. As recognized in the maturity model theories, all organizations are at different levels of maturity and this idealistic Lego block approach will be a challenge to deliver. The author still can recall a sore memory of the past when it was necessary to repeatedly go to the shipping department and physically pack his completed product for delivery to meet the customer's contract schedule and thereby earn a significant revenue income. This is not the best of examples but shows the linkage between the organization and its project processes. Having a positive support culture is a major component of successful delivery.

In the commercial environment, the relationship between the sponsor/owner and the project needs to be tightly coupled. A project is technically being performed to satisfy some organizational need. The primary relationship between these two entities is as sponsor/provider, with the project organization supplying resources and therefore also being labeled the provider. In this paired relationship the organization has an implicit role of supporter. Even though the project team produces visible work, the organization's support level can make the difference between success and failure. Absent this, the project team would have to create a complete support structure consisting of facilities, needed processes, and other operational functions.

The examples above provide good general views of the support process but as usual, the DoD level of formality shows what might be approached in a large organization. One can use the specification outlined below for cost and schedule-oriented processes needed in traditional project support. There are numerous examples of specific support areas relevant to the project's needs. A sample list of these is:

- Financial accounting system
- Procurement system
- Payroll and benefits system
- IT Infrastructure
- Enterprise software (including project related)
- Physical facilities
- Status reporting system
- Change control process

DoD Operational Support Model

As described in earlier chapters, the DoD has been a leader in describing various aspects of the project environment, necessitated by their third-party vendor environment. An initiative spawned in the 1960s attempted to define and standard-ize various processes related to project control. In this case, the sponsor was DoD and the supplier was their contract vendor, but the type of relationship fits our topic. This specification, originally titled DODI 5000.1, was released in the late 1960s for vendor comment. Not surprisingly, very few of the vendors could satisfy the breadth of these process specifications. For the next 50 years, this vendor standard-ization goal has been evolving and various standards documents still exist related to it. Somewhat surprisingly, the original 32 defined standard project processes have remained as part of the current guidelines. In the industry vernacular, this specifica-tion was known as C/SCSC (Cost/Schedule Control Systems Criteria), and threads of it now appear in various specification documents. The long-term stability of these specifications for a waterfall-type project structure gives credibility to the complete-ness of this process overview definition for that area. As the specification exists today, it is focused on generating standard Earned Value project status calculations but the architecture of this specification is more general than this. It represents the project control architecture that a mature organization should have in place to support the general cost/schedule management processes of a project. In any case, this specifica-tion offers a good road map for organizational process control architecture. Based on its long history, this specification represents the "Good Housekeeping Seal of Approval" for organizational systems to support project planning and control needs.

In the DoD and broader governmental sphere, this specification has stimu-lated vendors to improve their overall planning and control process; however, it is still a challenge to be certified as compliant. A litany of three-letter acronyms has emerged from this effort along with a sophisticated management control process. First, the concept of an integrated baseline review became a vendor requirement. The purpose of this review was to validate that all of the necessary project elements had been considered and were represented in the plan (NASA, 2019). After success-fully passing this review phase, the project is then baselined for control purposes. This control point is then used to evaluate the expansion of the approved scope. Reviewing back to an earlier discussion regarding techniques to handle scope change, it is interesting to see that task padding is not admitted here. Given the model principles outlined thus far, one can see that task padding destroys the valid-ity of control as defined in this specification. In other words, it assumes no padding but that is not the actual culture. A second standard metric that has emerged from this stimulus is the use of Earned Value (EV) parameters for status analysis. Some supporters will claim that this is the best status metric available, while others argue that it can be manipulated and is therefore not worthy (i.e., task padding again). The author's opinion on this is that the metric is as advertised if the underlying process architecture is as described in ANSI-748 (see below) and task estimates represent legitimate unpadded goals. There is one philosophically debated topic related to this line of management. That is, "is the level of management visibility and control needed by a commercial organization worthy of the administrative effort needed to produce this granularity level of status?" More discussion related to the proper level of management visibility will come later after the integrated model is described, but this is a tough question. Regardless of the answer, the data-related process structure outlined in this specification should be a good checklist guide for examining planning and control needs.

ANSI-748 Processes

The current title for the original specification outlined above is now ANSI-748 and it contains the original 32 items grouped into five major process groups. These are:

1. Organization (five guidelines)—Defines contractual effort and assigns responsibilities for the work
2. Planning, Scheduling, and Budgeting (ten guidelines)—Related to Plan, scheduling, budgeting, and work authorization
3. Accounting Considerations (six guidelines)—Accumulate the cost of work and material. Report on progress/accomplishments to date
4. Analysis & Management Reports (six guidelines)—Compare planned, earned, and actual costs, analyze variances, and develop estimates of final costs
5. Revisions and Data Maintenance (five guidelines)—Incorporate internal and external changes

The 32 detailed specifications are scattered among the groups as shown in the list above. It should be obvious from these high-level titles that much of the specification is focused on cost and schedule control. Another control focus is on "work authorization," which deals with formally controlling scope change and its associated budget impact. Realize that the class of projects being managed here is often in the billion-dollar category. What is not so evident from this list is the level of data integration defined across the groups.

There are some clear design goals embedded in this architecture. Some of the most notable are:

■ Using the WBS to link both the defined work and the organizational unit involved
■ Using a formal definition of Control Accounts to organize cost planning and control
■ Definition of a Performance Measurement Baseline for status comparison (plan versus actual)
■ Extensive budget tracking, including a management reserve

As indicated, this set of specifications was conceived in a period of trying to mature the definition of a waterfall delivery model and it still shows that bias. It does not fit the iterative model well but some common elements could be applied.

All projects are under some level of scrutiny by the financial and senior management levels. One of the common sayings about the agile method is the project is over when the money runs out. This is a not-so-subtle slam in that there is little schedule or cost control process defined, primarily just iterative sprints. The theory of iteration suggests that this process is over when the customer is satisfied, but in reality, financial constraints may ultimately define the stopping point regardless of customer status. The Critical Chain approach fits pretty well into the predictive data collection structure, even though the interpretation of project status would be different from what is outlined here given the modified task estimate approach. The CC process would create more task overruns and this has implications for a traditional status interpretation.

Yogi Berra the great Yankee baseball catcher sage once philosophized "if you don't know where you are going, you will end up someplace else." To help ensure that this is not the result of a poor project environment, a support structure as outlined here is the mechanical linkage between the organization's vision layer and the project team who executes that vision. Organizational process support architectures such as the ANSI-748 specification set offers a data and process blueprint for organizations to support their project portfolio environment. In addition to this formal list, there are many other support functions and processes that need to be supplied. See Richardson and Jackson, chapter 34 for more details on this (Richardson and Jackson, 2019).

The Model Support Structure

The next 20 years promise to produce a significant level of process change and a similar increase in support-related technology in the project environment, and it is hard to anticipate exactly what new support requirements might arise over this period but one of the most obvious trends is in the increasing impact of technology on project activities. The organization's information architecture will play an increasing role in supplying various communications links among the various project players and offering *Google-like* data access. At the far end of the spectrum, one might envision the availability of all project data being accessible in real time for all. This would include all of the artifacts of current and past projects. Communication among the global players will still be a need, but the method of delivering that will continue to morph with technological changes. Previous research studies have concluded that communication gaps are the leading root cause of project failures. Accepting that, the availability of needed data and personal communication availability becomes a key requirement. The evolution of cell phone "Face Time" has already introduced a new level of personal communication, and there is no reason to believe that even more robust tools will emerge. Future advances in this area will be prime candidates for improving organization, stakeholder, and team interconnection.

The next chapter will move further with another key supporting function. That is, how to decide what projects should be approved and the associated processes that go with that requirement. We will end this high-level philosophical overview with one more metaphoric memory example. The ideal case for the project is to look at the relationship between it and the host as a Christmas tree and an ornament. The project is the ornament. If all goes as desired, the project ornament can simply attach itself to the tree with a standard hook. Hopefully, the metaphor examples used here make sense; they are good memory triggers for the sponsor/project relationship.

Summary

A review of the current project environment does not show clear recognition of this view. Logic says that what is described here is valid. If there is less recognition than should be occurring, here are some possible reasons for this gap:

- Project execution is not being done in a standard form
- Organizational status systems are ad hoc and non-standard
- Senior management does not see the need for formalizing this sort of thing
- Technical project managers believe that they can't change the organization
- The organization is too busy to improve internal processes
- There is no recognition of the value of this approach

Regardless of the reason for not pursuing this goal, it represents a viable strategy to improve delivery results and decrease internal project costs.

References

Farrokh, J. and A. K. Mansur. 2013. Project Management Maturity Models and Organizational Project Management Maturity Model (OPM3): A Critical Morphological Evaluation. *Project Management*, 2(7), 23–33.

Ibbs, W. C. and J. Reginato. 2002. Measuring the Strategic Value of Project Management, Presented at the Impresario of the Construction Industry Symposium, Hong Kong.

Masters, S. and C. Bothwell. 1995. CMM Appraisal Framework, Version 1.0. CMU/SEI-95-TR-001. Software Engineering Institute, Carnegie Mellon University. http://resources.sei.cmu.edu/library/asset-view.cfm?AssetID=12323 (Accessed July 15, 2022).

NASA. 2019. Integrated Baseline Review (IBR) Handbook. https://hgtrs.nasa.gov/api//citations/20200000302/downloads/20200000302.pdf.

Richardson, G. L. and B. M. Jackson. 2019. *Project Management Theory and Into Practice*. 3rd ed. Boca Raton, FL: CRC Press.

Chapter 12

Portfolio Management

Introduction

One of the operational issues of an EPMO function is its ability to accurately forecast the value of a project proposal. The point is made in this chapter that a poor selection of a project represents a failure just as much as poor execution does. Given that organizational resources are connected across both proposed and current projects there is a link between the tactical level and the strategic. Because of this, the two functions should be viewed as an integrated whole.

Another view of this topic involves how project proposals also represent different types of decision needs. These are titled RGT decisions for Run, Grow, and Transforming the organization. Each of these types has a different perspective regarding Return-on-Investment calculations.

The final section of the chapter offers different implementation options with warnings related to how each might be accepted by lower levels. It is emphasized that acceptance of the function has to do with the perception of value by the lower levels. Regardless, implementation of the function lies in the hands of senior leadership who has to accept their role in this process.

The final segment of the integrated project delivery model involves the concept of a project portfolio. At first view, this seems to be a self-evident idea, yet in practice often fails to achieve the desired goal. As the practice of using the project model to deliver organizational changes became more popular, it became more obvious that the various project initiatives were consuming the same limited resources. This highlighted the need to have a formal selection process to pick the best options and the informal approach was not working because of local bias. If you have limited funds and need a new car and the house painted, which do you do first? That is the essence of the portfolio process. One of the best visual examples of this problem can be simulated by the scenario of inheriting $100,000 from Uncle Joe. Where should

I spend or invest these funds? What are my goals for this money? How much risk am I willing to take? What are the steps to make these decisions?

In looking at this situation, there are zero arguments to refute the notion that some formal process is needed to deal with this issue. That recognition was the stimulus to create a function often titled the Project Management Office (PMO). The early primary reason for creating this organization was to help manage the selection of projects that would optimize organizational goals. Selecting the wrong project would waste resources and potentially omit acceptance of a better option. The fundamental PMO goal is clear, but the implementation of this functions at high organizational levels is fraught with political and leadership issues. The integrated model cannot fix this internal conflict issue but can make the need for such a requirement more visible.

In a PMI 2017 survey, 70% of the organizations claimed to have a PMO (PMI, 2027). However, 50% of PMOs are shut down within three years for various reasons, some of which are outlined here. Ninety percent of the senior executive surveyed believed that "activities" that support strategic goals are vital to success. Note that this positive response does not explicitly say that a PMO organization is required, but does support a high-level vision-type process. Lower project levels often perceive functions at this level as taking away aspects of their role and that may well be true.

Organizational Design Issues

It is not surprising to find that an organization whose role is to select among competing options would not be viewed favorably by some. That is likely the fundamental case with this function. The typical name for this function is PMO and it remains a popular organizational entity, albeit with varying roles. While statistical data does not specifically uncover the root cause of this less-than-desired result, there are two basic conflicts at play. First, the lower project level advocates do not like another decision source between them and senior management evaluating their merit and selecting alternatives. In the industry vernacular, this is often described as "calling my baby ugly." Project sponsors are very sensitive to such. The second potential source for variation comes at the sponsoring higher level. Senior executives often do not document the defined goals of the organization that are needed to drive the PMO, nor do they see that active leadership of the process is their responsibility. In addition to this, the C-level may desire to be cutting-edge without understanding the complexity of such a goal. The real key to PMO success comes from active and supportive leadership. An operative vocabulary term for this role is *alignment,* meaning the connection of project initiatives to business strategy. The thesis here is that organizations will not automatically achieve alignment without some type of active formal structure to support the project decision process.

EPMO Model

Beyond the organizational relationship aspects outlined above, there is also a breadth of scope issue to consider. Assuming the basic theory of this function makes sense, it should be applied across the total organization and not by a department. PMOs have historically been linked more to the IT function than others, but this is also common practice in making capital spending decisions.

Formalizing this function across the organization is the proper strategy and a typical title for such a function is Enterprise Project Management Office (EPMO) (CIO). Even though the fundamental role of each structure is similar, the EPMO model focuses more on a broad organizational strategic view of projects. Concepts such as standardized processes, governance, and best practices are emphasized. Both the PMO and EPMO organizational types are judged to be, philosophically, technically, and politically operationally complex. The overall potential rewards are higher with the broader EPMO approach, but the added implementation complexity seems to inhibit some organizations that choose to leave the function as a department-level PMO.

Decision Threads

The process of defining a project is to first have target areas or goals outlined at senior levels of the organization. An example of this is "we need to improve our customer service." That doesn't say how to do this but gives guidance. From that high-level driver, various initiatives can be defined and quantified by the PPM as to development factors (time, cost, resources, benefits). The potential view of a project begins to emerge with this. One of the tools for looking at project proposals is a "bubble" chart. A hypothetical sample is shown in Figure 12.1. The proposals are arrayed across the cost versus benefit axis. This view provides a visual method for seeing how the proposals can be ranked. In theory, the selection decision would be to take the highest benefit-to-cost initiative, so the bubble in the top left corner

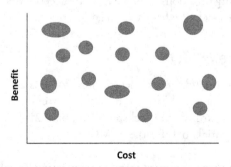

Figure 12.1 Project bubble chart

would be the highest-value candidate for selection. Bubble sizes can also be used to show the resource size of the project.

But these are just some of the decision criteria used in the selection process. Not all selection decisions can be made purely based on a raw cost-benefit evaluation. High-value large projects can be broken into more manageable smaller groupings. And three additional business views need to be considered. These are:

Running the business (R)—keep the doors open (current operational mode)
Growing the business (G)—improve the current business (expanding current model)
Transforming the business (T)—moving to a new business model (major risk but the high potential reward).

This decision perspective is called the RGT. As one can imagine, picking the right project targets across these views is very complex. Transform decisions tend to be long-term with low visible payback and high risk. There has been a great deal of discussion in the industry regarding the proper mix of these three groups. As an example, some organizations suggest a project resource expenditure mix of 70, 25, and 5, respectively. Higher-risk organizations such as Apple and Amazon could point to their improved competitive position by looking at more transformation-level decisions. One can only imagine the meeting at Apple many years ago when the proposal was made to move business focus into a small music box (iPod) and later morph that into a cell phone device. The first of these was not strategically successful but the second transformed the company. Both of these decisions could have taken the company toward bankruptcy. This class of decision is called *disruptive transformation*. Many times, this level of direction change can make or break the organization and therefore must be done carefully. Four successful contemporary examples of the Transformation type of decision are Amazon Prime, Airbnb, Uber, and Google Maps. Only the Google and Amazon decisions could be somewhat classified as a major "grow" initiative that could be started with relatively smaller resources and iteratively expanded after customer acceptance of a new business option. Large organizations struggle to find success stories of this magnitude, which may indicate a lack of effective leadership.

Functions of the EPMO

From this point on, the text model descriptions for this centralized project decision structure will be called EPMO, even though current industry practice is heavily named PMO. Richardson and Jackson (2019, pp. 521–523) offer the following summary list of potential goals for the EPMO:

1. Strategic alignment—support formal organizational goals
2. Resource management focus—matching and managing resources to projects

3. Project control governance—this can take on a control focus based on schedule and budget
4. Efficiency—this relates to a management control view and one most disliked by the project teams
5. Balance—oversight of the RGT mix
6. Value optimization—this function attempts to evaluate the operational value of the project, both at approval time and during execution

It is important to understand that the EPMO goal list for a specific organization may well vary from above based on what the organization chooses to achieve. Grey Campus describes the following three models for a PMO and we add an extra fourth (Grey Campus):

1. Supportive—Acts as a repository of project artifacts
2. Controlling—Primary role is an auditor of projects
3. Directive—Have a high degree of control over the selection and management process
4. Creating Centers of Excellence—Housing subject matter experts in the PMO to assist projects; this approach could be embedded in any of the three basic models

Realize that there is little standardization of EPMO functions, so these examples only provide a high-level overview of philosophies. The only core responsibility across all design models is to support the organizational project portfolio decision process, which may include all of the functions outlined above or only a skeleton set.

Functional Responsibilities

Alexander (CIO, June 6, 2018) offers the following four key EPMO roles:

1. Prioritize projects and programs
2. Evaluating resource capacity
3. Project risk assessment
4. Monitoring the status of ongoing initiatives

The degree to which an organization chooses to allocate roles and decisions to this function remains varied. In addition to the more straightforward operational functions listed above, other worthwhile roles may be assigned here. This list includes:

1. Assisting management in the development of formal organizational goals
2. Defining project structures into phases or program groups

3. Providing technical help to the project layer—i.e., estimating, training, consulting, tools, etc.
4. Assisting in developing a formal governance system to standardize organizational authority relationships

All of these various role functions listed are conceptually valid and fit the concept of an EPMO; however, like so many things in organizational theory, centralized control is not always the best answer. A stifling bureaucracy can be created by this activity, which would negate the theoretical value. When a centralized function can be perceived to be a support helper at the project level, it will be more readily accepted. Also, participation in the decision process by the lower-level entities would help with it being accepted. The adage "I am from headquarters and here to help" can be taken two ways. Similarly, management's perception would be viewed positively when the project decision quality is credited to the EPMO function's involvement. This function is often viewed as overhead and suffers from the same level of acceptance as found in other similar activities. There is one role that will result in the universal rejection of this function. That is, charging the project for any of its services. It would be exceedingly rare to have any high-level function charge that was willingly embraced at the project level and this is no exception.

Linkage Relationships

The EPMO concept represents the project birthing link in the integrated model component group. This is not to be viewed as some activity separate from the project level. Data needs to flow up and down the organizational chain between the active project and strategy levels. The resources involved in active projects are often the same resources that could be allocated to new projects, so there must be a global evaluation of ongoing efforts. One of the toughest decisions that an organization can make is to decide that an active project needs to be terminated. Project managers are notorious for believing that they can recover from a failing effort, but because of that trait, this decision needs to be held at a higher level in the organizational structure.

Regardless of the local project assessment approach, there needs to be an appropriate level decision made regarding the following:

■ Unbiased value assessment of the proposed initiative
■ Requirements or features to be produced (graded by required and nice to have)
■ Schedule and budget estimates (to not exceed constraint levels)
■ Appropriate level risk assessment
■ Defining status tracking requirements (not all will necessarily be the same)

An important step in this process is an organizational discipline to ensure that no projects will be authorized without formal approval at the EPMO level. Each proposed and approved project will be recorded in the EPMO tracking database. Finally, for a project to be funded and resources charged, a formally approved Charter will be required. The Charter should be the mechanism to initiate the ability to approve some budget level and allow charges to that project code. If deemed necessary, an exploratory Charter can be approved to develop an expanded understanding of the target area. The EPMO function can be looked at as a dual-headed activity. The decision-making layer is the management portion and a supporting technical activity that is responsible for the data required to evaluate this complex decision process. This support function is known as Project Portfolio Management (PPM). For this discussion, visualize it as a data and support function between the project and EPMO structure. A formal Charter needs to be the mechanism to reflect the status of a project. No Charter, no project! Expenditure of organizational resources is the responsibility of management and the authority to approve a Charter is a reflection of that delegation level. From a theoretical standpoint, this should be from one group and that is the EPMO. Any operational weakness in this resource allocation decision drains off the availability of those resources that could be used to better fit organizational goal alignment. This selection process is the one that causes the function to be most disliked.

Wrap Up

There are many examples in the project world where some high-level decisions gave an organization a competitive advantage. These can be found in organizations such as Apple, Toyota, Honda, Amazon, Walmart, and many others. Not all of these were made through an EPMO organization but they were made through a management level that had a creative strategic vision. It is much easier to find project decisions that lacked such creativity. The cardinal management rule to remember here is that a bad project decision cannot overrule good project management even if that poorly selected project is completed successfully according to traditional measures.

Project selection is more complex than just approving the project with the highest payback. The RGT concept equation is often not followed. If an organization spends all of its resources improving the Run portion and never deals with longer-term ventures, it may well find itself out of business as the market environment moves. The Grow option is often recognized but growing too fast can also cause problems if the current Run state is affected. Finally, the Big Elephant is to find that thing that the world wants—Apple iPhone, Toyota Prius, Amazon, Prime, Disneyworld. Each of these strategic decisions transformed the organization, but each could also have been a major failure. Transforming the organization is the most difficult of all and may not even be the right concept for some organizations. This is the fuzzy world of the EPMO. It is interesting to watch organizations as they move through their life cycles and wonder how this class of decision was made. The EPMO function represents organizational survival over the long term and for that reason must be tightly connected to the project environment. When you deal with an organization where it is impossible to contact a human or wait on the phone for 30 minutes to talk to someone whom you can't understand, what reaction do you feel toward that organization? This is an EPMO-type Run question. What might happen when a well-run organization enters the market without this reputation? How long can bad management exist?

It is time to take this theory and start combining the puzzle to assemble the integrated model.

Hopefully, this overview of the strategic side of project management has sensitized you to the role it must play for success. This decision arena is a mandatory component of this problem and is not a separate function as currently structured.

References

CIO. 2022. What is an EPMO? The Organizational Key to Project Success, CIO Foundry.

Grey Campus. 2022. Which PMO Structure Is Right for Your Organization? https://www.greycampus.com/blog/project-management/which-pmo-structure-is-right-for-your-organization (Accessed December 5, 2022).

PMI. 2017. *Pulse of the Profession 2017.* Newtown Square, PA: Project Management Institute.

Richardson, G. L. and B. M. Jackson. 2019. *Project Management Theory and Into Practice.* 3rd ed. Boca Raton, Fl: CRC Press.

Chapter 13

Integrated Model
Design Components

Introduction

The goal of this chapter is to begin compiling the necessary components and work processes needed for the new model. Data for this is derived from various fragments uncovered in previous chapter descriptions of the classic models that have attributes suitable for inclusion in the new integrated model. From this, a skeleton decision architecture will be outlined that includes multiple concurrent work options. This is the most difficult chapter in the text because it combines various pieces that do not naturally mesh into a more global process for dealing with what has been described as success-oriented processes. This description is intended to rationalize the core skeleton of the new model.

Two topic areas will be excluded from the model description. These are HR and quality theory which are both assumed to be embedded in the internal decision logic. The HR component is a vital part of success but technically is represented by the project team and not the model itself. Second, the quality area is also important in a conceptual sense but is considered to be represented by the internal culture of the project team and support organization. Both of these topics need to be understood and utilized as a part of the management culture.

Chapters 6 through 12 described various delivery components that have identified incremental value items that have the potential to improve project outcomes if they can be integrated reasonably with the project profile goals. As stated multiple times, no single one of the classic models fits the design criteria, but each has interesting characteristics that may be useful for specific functionality—i.e., speed, control, user satisfaction, risk management, etc. The remaining challenge is to position

DOI: 10.1201/9781003431091-15

the decision groups for an appropriate delivery strategy. At this point, we have some useful tools but now need to work on summarizing the processes that will become the model Lego blocks.

Selecting Major Model Drivers

Outlining the model structure is the key step in making the overall process logic fit the various success drivers previously defined. All management models need a logical and understandable structure and this one is no exception. It seems logical to place this decision location at the highest level in the structure a strategic driver for project selection that initiates the development life cycle. This is an expansion of the traditional view which does not explicitly recognize this step as being integrated with the development view. From this starting point, the development process is envisioned to be a series of elaboration decisions. This sequential starting decision is represented by a vertical group of steps that seem to be more representative of the elaboration idea, rather than a time-based traditional view. This is now modeled as a top-down layered process. The following five decision groups are packaged into the layered structure:

1. A group that contains major decision processes covering the entire life cycle.
2. A group that supports techniques to structure the project into an appropriate delivery structure.
3. A flexible work execution structure that supports the requirements defined in the project profile and also allows flexibility in the level of formality for various processes.
4. A work management process that facilitates different forms of work within the structure.
5. A group of flexible communication techniques adapted to fit the profile goals.

Weaving all of these groups into one understandable structure is the challenge of the next chapter. Five high-level requirements are key success drivers of the integrated model design. These are summarized below as key gaps that are often linked to poor delivery results and are therefore appropriate processes to monitor (note brackets at the end of each item):

1. A visible management gap exists in the decision process that spawns project approval within a competing portfolio. Organizations must have a global perspective and rank projects according to value. (*Portfolio management*)
2. Poor requirements definition is related to stakeholders being tentative about being involved with the project, thinking that it is either too technical or they don't believe that this is their role. (*Requirements definition*)

3. Resources are often not managed well in terms of timely availability or skill. Agile tends to resolve this by having a dedicated sprint workforce. Schedules automatically slip when resources are not available on schedule. (*Resource management*)

4. One of the greatest evils in the traditional model is the padding of time estimates and the culture of procrastination and multitasking. Note that this is an implicit attribute of any formal model but is not a common operational practice. (*Task or work estimating culture*)

5. Budget creation, scope change management, and completion forecasts in the traditional model are error-prone. These items are associated with model assumption errors. (*Scope management*)

At the highest level, these are the primary targets we are attempting to deal with and at least highlight some of the key success roadblocks that have been identified. One must be sensitive to the notion that these are the WHAT items to resolve and more needs to be understood regarding the HOW aspect. Fifty years from now there will be new technology-related answers to this set of questions. By then, projects may be driven by simply telling an automated robot what you want. For now, we see this aggregation of work, process, and architecture as the key targets to pursue.

The next element to include is additional logic related to the work management mechanics regarding how to decide which delivery option to select. Each work execution option has both positive and negative attributes so the challenge is to understand the tradeoffs for each of them.

Four validated task execution models are the prime vehicles to produce the desired outcomes. These are predictive, agile, Scrum, and Critical Chain. The execution layer is supported by higher levels of management decision process layers. The highest management layer includes senior management roles in the strategic visioning and planning process. This top-level group is required to be an active player in the decision-making process, and it will be closely linked to the portfolio analysis process that produces data related to the proposed projects. This relationship formally releases approved projects to the next decision layer. Each subsequent layer adds key decisions to the process based on a cascading information flow. The formal term for sequential steps like this is *elaboration*.

Let's digress here for a moment and use a simple memory example to illustrate what the elaboration process entails. Assume that senior management, after diligent study of competitive data, decided that the best strategy for the organization is to produce a hypothetical device that "Leaped tall buildings with a single bound." This project would be approved based on this requirement along with possibly some deliverable constraints based on time or budget. The example that follows represents the elaboration issue. If you were the project manager for this scenario, can you think of a few questions that need to be answered before moving into execution? That would be step two in the process. Here are some sample elaboration questions that need to be resolved for the next layer:

How tall is the building?

What are the goals regarding cost?

What are the constraints that can affect how the project should be planned?

This line of questioning should go on until a clear understanding of the general project deliverable function goals is understood. Oftentimes, senior management and users believe the original goal statement is sufficient to start project work, and this example shows that is not the case. A poorly defined scope is the beginning of project failure, similar to approving the wrong project. In some cases, decisions made at a lower level need to be communicated to or approved by higher levels. In the case of scope elaboration, some details need to be approved since this specification will be the guidance used by the technical team to produce the outcome. They may say that it has to leap at least a ten-story building in no more than two tries and the unit cost must be $X. If questions such as these do not get resolved early in the process, they become potential failure points later. Also, it is important to pass on to the next step any constraints that could affect delivery results. In like fashion, lower-level decisions need to be visible to high levels.

A follow-on elaboration step is concerned with macro-level issues related to the project design packaging. For example, should the project be broken into phases to complete certain requirements earlier? Typical packaging could be a priority of phases with design functionalities, or it could be a packaging of key skill groups as subprojects such as hardware, software, and networks.

The third layer involves more specific deliverable planning related to the actual execution work strategy. A WBS is a popular method to describe various structures. This layer is the transition from a logical project view (WHAT) to a physical (HOW). At the bottom of the elaboration process, the work units are defined by one of the four task execution delivery options.

The decision process envisioned above can be viewed like peeling an onion, with each layer providing the insight needed for the next. This is the management process that takes the fuzzy requirements and answers the string of detailed questions needed to eventually decide how best to execute the needed tasks to satisfy the delivery goals. This type of process is often viewed as going from a logical view to a physical one, or going from the what to the how. Across all of the decision layers, a robust communication process must be in place to make decision-related data freely available.

Model Structure Elements

A significant breadth of project management theory has been described throughout the previous chapters. It now seems timely to begin compiling how some of the process units can be part of a structure for use in an integrated model. The defined goal of this model is to execute work units based on multiple delivery requirements dictated by the project's characteristics and goals. The previous background research uncovered multiple processes and components that need to be included in the new model. The following list summarizes some of the key decision processes required in the model structure:

1. A process to select the project from a portfolio of candidates
2. A process to profile the project characteristics and delivery goals
3. A process to package the project into macro-level groupings such as phases, subsystems, or other collections
4. A decision mechanism to define the type of work that best satisfies the delivery goals

Beyond the model decision structure, there is also a need to recognize best practices process gaps found in the current models and make sure they are dealt with. The following is a selected list of these:

1. One of the most visible and important early decisions is to select a project that aligns with the organization's goals. This decision process needs to be a strategic function in the organization and was defined as an EPMO in Chapter 12. The process should be part of the integrated model and not be viewed as a separate activity.
2. Project stakeholders are often tentative about being actively involved with the project, thinking that it is either too technical or they don't believe that this is their role. This omission from the process will lead to an inferior result, or even doom a project. Projects should be terminated if there is no appropriate user involvement.
3. Resources are often not managed well in terms of timely availability or appropriate skill. The agile and Critical Chain models have shown the value of having a dedicated and focused workforce. It is important to recognize that schedules automatically slip when resources are unavailable on schedule, yet this is a commonly observed event. Availability of required resources on time is one of the most important management criteria for success.
4. One of the traditional model band-aids involves the padding of task time estimates to cover variances in estimates. The Critical Chain model discussion in Chapter 10 showed how this practice makes the overrun worse.

In addition to this itemized reminder list of common high-impact gaps, there were various best practices outlined in the various model discussions. These are assumed to be the norm in the integrated model.

Project budgeting and scope change concepts in the traditional waterfall model are error-prone. This model process does not offer an accurate way to reflect the approved scope change and risk situation. In addition, most organizations do not properly handle specific variances created by task overruns, risk events, or scope changes. Many common practices related to these areas are hidden by hiding their impact through task padding and this distorts accurate views of the project status.

The term elaboration has been used to highlight how project decisions are made, starting with a fuzzy vision and then step-by-step moving to the physical execution step. The core layered decision elements summarized above serve as the backbone structure for the integrated model architecture and represent its design skeleton. In addition, the important roles of resources and organization process support (Chapter 11) also need to be reflected in the model.

Work Management Strategy

An early view of this effort was to define methods to have multiple work queues based on predictive and iterative goals. Digging deeper into that goal produced the recognition that the project characteristics should be the driving factor for work methods, and this caused a major change in the design scope for the effort. Multiple techniques were examined in previous chapters, and each was found to have some relatively discrete positive attribute that gave it the potential to be used in an integrated model with various work delivery options. Each of these needs to be understood in their defined decision block. Also, we need to keep in mind that this discussion still has some characteristics of the six blind men story. Collectively we need to be sure to define the whole elephant. The sections below will offer brief comments to start filling in the execution level and support contributors that fit into the overall puzzle.

Using the Waterfall Model

Ah, our old friend: This is a well-known simple management structure. Requirements are faithfully documented, and there is a fixed definition of tasks and timing. Associated resources can be defined. The formal plan contains items that management wants to see such as the completion date and cost. A risk assessment process is defined. Management has everything they perceive to be needed. The model is management nirvana! However, we have identified one little problem. This view often does not fit the project characteristics that it is being used for and it often is not performed per the design. Band-aids are often applied to produce the desired views, but they often are worse than the problem. Task overruns can be fixed by adding extra time or buffers, but management does not like buffers. There are situations where it is wasteful to spend so much time producing planning documents that have no operational value. Regardless of these recognized shortcomings, the waterfall model visual view of a project is the best one available. If the project's profile and delivery goals fit the model, it is a very mature view for task overview and perceived status tracking. Definitional items such as the WBS, network calculations, and Gantt charts have survived through the years and are worthy tools for the future. In addition, the following defined organizational processes are linked to this model and have universal value for all projects:

- Portfolio analysis
- Business case (justification)
- Management approval (control)
- Requirements specification (scope)
- Risk assessment
- Change control
- Formal status reporting

Seventy years of evolution make the waterfall view the most mature project management environmental specification. If execution speed is not the main requirement but rather overall control and risk management, this may well be the best management strategy.

If one were to attempt to "tweak" this model to accomplish alternative goals, one strategy would be to decrease the front-end analysis detail level and leave more of the lower detail specifications to be filled in at the task execution level. Formal documentation execution speed could be improved through the use of templates, previous project data, and improved information technology. Active involvement of users is a traditionally defined waterfall goal but generally can be improved.

There were previously described examples of techniques where iteration and Critical Chain techniques can be embedded in the waterfall structure to improve output. The key point of this is to recognize that the waterfall structure has known value and known weaknesses. Both agile and Critical Chain concepts can be introduced into that structure without destroying it completely and proper use of best management practices is needed to improve successful outcomes. The waterfall model represents a reasonable starting place for managing a project so long as one is cognizant of gaps between the project profile and the model assumptions as outlined in Chapter 8.

Evaluating *the Iterative Model*

No alternative management idea has been more widely accepted in the project community than the agile approach to delivery. Various dialects have also spawned, notably Scrum, with each bringing new vocabulary, tools, and modified processes. There is sufficient survey evidence that this method creates higher user satisfaction than traditional approaches, but there is also suspicion that this looser approach will not be accepted by many management cultures. Owing to its touted success, it is impolite to criticize the model, but some subtleties need to be understood. On the positive side, agile has taught the industry a valuable lesson in project management delivery improvement, even though one might have a hard time describing exactly what that is. First off, the defining principles of agile are more motherhood than specific so that is not it. Second, from the 12 principles described, approximately seven of those are previously recognized as good management practices for any environment. Yes, it is good to have an active user evaluating output and making corrections for another pass, but what if the product cannot be torn down to start over again software can? The remaining principles are hard to strictly credit to agile. Probably the most noticeable productivity aspect of an agile project is the execution (sprint) structure that this model pioneered. It encapsulates how the work is defined within the team with user involvement and has discipline regarding daily communication. All of these elements contribute to the positive result. Sprint-defined tasks are matched to available resources which are also often not followed in the traditional project but clearly defined in the associated agile sprint mechanics. Agile has an implied speed focus on tasks given that the sprint is a fixed-time block. In other words, get as much done in the fixed-time sprint as possible. Status reporting is mostly for the team. Predicting long-term output is not as clearly defined as in the traditional model. One of the more arguable positive aspects of agile is the feeling that the project team with user support has been delegated more control over the actual deliverables.

If the project target were building a bridge, one can see issues in how a looser task definition would work. That said, there are lower-level decisions in almost every project environment that could be executed using agile-type principles. One must ignore whether you agree with agile fully or not at all. Recognize that the method has been very successful in the IT delivery environment and try to utilize those aspects for other project types. It is also important to recognize that some of the agile principles do not fit well in tangible product situations. One cannot just say that the model doesn't fit, therefore ignore it. This absolute rejection attitude does not fit the learning organization theory. The following list contains six agile-style management characteristics that need to be recognized as having improved productivity value in all projects:

1. Producing leaner requirements based on the project profile
2. Streamlined documentation

Professional Management Standards

There are numerous sources to find standard project management theory. Sample well-known examples are PMI's PMBOK, DoD specifications, and UK's PRINCE2. In addition, there are many books outlining the broad view of management theory. Collectively, sources such as these have contributed to the understanding of key required management processes. By definition, most of these are broad theories and do not undertake prescriptions to describe how to execute the defined elements in a specific project environment, nor do they focus on how this varies by project type. These theory-related sources offer a good understanding of the overall process but are less effective because they are focused on broad industry views. This makes specific help less feasible. This text is freer to be critical of current practices since the design goal is to map out an improved direction. Because of the broad theoretical perspective, these sources do not get into the negative aspects of the various theories and this is needed for the new model goal. Think of the various industry-vetted literature such as the PMBOK is an important element in the search for the holy grail of project management. Academics would call sources such as this Common Body of Knowledge. After being exposed to these formal theory sources, one is left with the state of still not knowing how to manage a project using such concepts. Most current training programs focus on the use of a single model such as waterfall or agile, more than attempting the broad view described here. The ideal knowledge transfer process would be to produce a specific team training experience for a local project environment. As a successful example, Richardson and Jackson (2019, pp. 278–283) described such a training approach called TPS that was sponsored by the Software Engineering Institute (SEI). The key point here is that a customized management process as outlined there should be taught to the project team.

Life Cycle Architecture

The best descriptions from the previous sections support the notion that multiple execution approaches are needed in the new model to fit the broader execution perspectives. The life cycle is now refocused on key success-oriented decisions rather than the traditional names for standard stages. The visual for this is to have each step building on the previous one, and needed details flowing downward from prior steps. The top decision area is the project birthing step and the lower area would represent actual work, and eventually, some concluding processes such as testing, implementation, and shutdown. Each block would be given a general title related to its function to represent the required decisions. From this, the core of the model would be a linked decision block diagram that will map the design steps and highlight the decision structure. The sections below will offer brief comments describing how other key success contributors need to be added to this physical structure.

Out of this fuzzy mist, the macro level of the management process becomes clearer. One of the new functions yet to be explored in detail has to do with the process of admitting multiple work options within the same structure. This is the major variation over traditional work views and one that requires more discussion. The traditional models did not have this function, so the selection of optional task execution methods was a non-decision in that regard.

This extra decision layer selects one or more of the four delivery options based on the work characteristics. Even though the first blush might lead one to believe that this is radical, it is more logical than the simplistic models in which the single execution option does not fit the project profile. This is the major logic variation from traditional models and is designed to better match the work goal. The primary difference here is the management process is dictated by the project characteristics from its inception and that link carries through the execution step and beyond. In this mode, the delivery approach shapes itself around the proper work execution requirement, rather than a more niche-oriented fixed task view as found in all classic models.

Previous chapters have attempted to show how different project characteristics and delivery goals have an impact on a fixed delivery method. We now see how the new project decision structure flows and that the project management view is broader than most currently consider. The definition of the project profile is now riving subsequent steps in designing the proper delivery strategy. There is recognition that some of the points raised here seem to challenge existing models; however, that is true only to the degree that some organizations use the wrong model for their project types.

Project managers are an intelligent and creative breed. In many cases, they have created their own unique way regarding how to deal with various gap issues. Some of what is shown here may go against that view. Admittedly, the approach outlined here involves a complex topic and the new model solution challenges several current practices. Hopefully, the various historical background chapters have been sufficient to make one at least accept the notion that there is a need for a revised management approach that includes multiple work execution options based on the design factors indicated. Although each of the classic models has some unique merit in a particular situation, their limited view of task execution suggests that none of them completely fits the design goal. However, each has some selected characteristic that needs to be included in the integrated view. Regardless of one's view of the classic models, it is important to understand which of their inherent techniques correlate with project goal alignment.

In order to achieve project success, one must understand the decision factors that drive the project to its desired completion. This requires a deep understanding of the various theoretical knowledge areas involved as well as the daily dynamics that can move the ongoing project in the wrong direction. The integrated model outlined here simply provides structure to the decisions required along with key best practices that are linked to improved outcomes.

Work Types Overview

One of the more significant management elements embedded in each of the classic models involves the way defined work is processed. When a task is executed in the wrong manner, gaps occur that often bring less desirable results. The following list represents the primary operational differences related to each of the target classic models:

1. Waterfall—tasks are arrayed in a defined linked network. Time estimates often include padding to cover time variances. Scope changes are not defined in the project plan, but additional work will be added upon approval (scope

change). Tracking status and control is a key activity. The plan resembles a bus schedule with fixed dates.

2. Agile—the work focus here is on producing outcomes based on loosely defined features inside of fixed-time sprint packages. User satisfaction is a key observed result. Control of the budget and overall schedule is limited to a macro level.

3. Modified Scrum—this is a dialect of agile that is designed for use with a more predictive scope environment. The new model is now describing this execution approach as a sprint-oriented general model. This execution option uses agile principles with a defined deliverable single sprint completion process. It is assumed that this work option will have to satisfy a MoSCoW-type scope definition at some minimal level.

4. Critical Chain—project work is viewed as a chain of defined tasks with time estimates set for 50/50 probability. The primary goal of this model is to execute the defined work chain as fast as possible. Resources are the key focus and buffers are used as the primary control variable.

The model segment represented by these four models is the optional work execution decision. Within a work environment, having a complex set of goals such as defined control variables, user satisfaction, and delivery speed does not fit the current classic models. It is more reality-based to look at work execution based on various factors, most notably the level of scope definition for the task involved.

Recognition of the need for multiple work options brings with it more sensitivity regarding the way in which scope is defined. A single-value approach found in the classic models does not fit reality, so the scope definition needs to be segmented into categories as defined previously using the MoSCoW concept. This gradation of the work deliverable requirement would allow a predictive Modified Scrum sprint option to become viable and this fits a typical characteristic of some work units. Defining flexible scope in this manner opens up the viability of using iterative sprints more frequently, even in predictive structures. The goal throughout the text has been to avoid inventing special vocabulary for the new model, but the concept of a modified predictive sprint work option requires an exception. A modified scrum sprint for a defined deliverable can be applied in a sprint structure with the caveat of "get as much of the graded requirements done as you can in a fixed timebox." This idea works so long as the graded scope concept is implemented and that seems to be another subtle recommendation for all situations.

Delivery Model Gaps

Various positives and negatives have been mentioned regarding the work execution options. The approach taken for the integrated model is to use only selected portions of the named models that fit the needs of the new design. One can emb more of the functionality from these models into the structure if they have a sufficient understanding of that additional piece. The selected portions referenced here are judged to have the most potential to match the flexible delivery need with minimal added complexity. In addition to the functionality of the work delivery options, it is important to also understand related model process gaps to avoid. Somewhat related to this idea is the various failure areas observed in industry surveys. We will now take one last critical sweep through the work delivery options to review what objective the method best fits.

Waterfall Model Analysis

If the target project scope is considered well-defined, a waterfall-oriented plan fits the project planning and control structure best of all options. However, one must look at history to see that this assumption is more questionable than it appears. To illustrate one simple example of this gap, imagine a well-defined task without proper resources available, even with a perfect estimate the plan will be in error; therefore, the model does not work as advertised. Yes, resource capacity management is part of the general theory but is often not well practiced. Second, the planning gap related to task estimating practice has been repeated here frequently. This practice essentially destroys the integrity of the model in both scheduling and control aspects. These two examples do not reflect the model being wrong, but the lack of understanding of the model's architecture and assumptions is the primary gap that decreases the model value.

One of the perceptions of the waterfall structure is that it provides desired control parameters through its fixed time task and cost calculations. The two examples above illustrate how the model does not reflect either practices or an understanding of the required mechanics. Management and stakeholders like the waterfall structure since it clearly describes the "anticipated" outcome of the project in terms of time and cost, yet in both cases that assumption is not valid. For example, assume the initial plan shows a completion date of June 4 and a budget of one million dollars. What happens when scope change, padded task syndrome, or resource management gaps occur? Each of these very common situations invalidates the goal of the model. For the waterfall structure to represent a more realistic management model, several additional processes need to be added, and even then, one can criticize the usefulness of the plan. Sloppiness in dealing with scope, risk, resources, and task estimating collectively makes the waterfall plan presentation questionable and often useless from a management standpoint. Without discipline in these areas, it cannot be recommended as a decision tool. Managing these processes represents reality gaps for the model. This is an example of a very neat model that does not represent the underlying set environment. So, the final grade for a model is to see that it matches work execution to the goal structure. These are the fundamental reasons why the model does not produce reliable forecast for project schedule or cost parameters.

Another somewhat behavioral issue related to the waterfall structure is its perceived role as a control model. The description above has shown just some of the reasons that it does not accurately reflect task or completion dates; therefore, it provides erroneous views of project status. Traditional management practice spends significant attention to planned versus actual comparisons as though that represents status, when in fact it doesn't. Calculation of these data items consumes considerable planning time, so there is questionable value in this early exercise from a status-tracking value view. Once again, the model is not wrong, it is just not representing reality. One must trace back to the operational processes related to scope, risk, task estimating, and resources to see where the reality gaps have occurred. The net result from this is to recognize that the nice time and cost factors calculated do not march reality well.

In comparing the use of the waterfall model for managing a project, the most negative statement that one finds is that it leads to a static work culture as compared to action-oriented as found in Agile and Critical Chain environments. The analogy used earlier is that it is like the slow city bus schedule. If the padded time estimate is for it to arrive at 8:15, there is no motivation for it to arrive sooner than that. Once again, this negative is not the model structure itself, but how it is used. In comparison, the Critical Chain theory has a completely different feeling of action. Also, agile and CC both have more realistic status-tracking views built into their reporting methods.

Waterfall status variances have been shown to reflect the wrong work culture and tend to focus corrective action on the wrong things. Despite these negative comments, the waterfall model offers the best potential project task architecture through its mature toolset. Recognize that this model has survived essentially intact for 70 years because it has been found to aid the management process, even though it is used improperly in many cases. Even so, the waterfall model offers a reasonable project overview and it can be modified to fit into the new model structure. That said, it could also produce more accurate results if the project team better understands the underlying assumptions and what other related variables needed to be included in the view.

Quality management programs of the 1970s taught an important operational lesson regarding the need to uncover problems rather than hide them. Waterfall tends to gloss over internal issues through excessive planning, padding of estimates, and having a single fixed work execution option, even though even casual analysis would show that these assumptions do not fit reality. In a classic quality case study, it was shown that carrying excessive inventory levels hid supply chain design problems and resulted in higher overall costs. This same operational scenario exists in many waterfall projects. Here is a specific example. If every task is padded to cover up risk, resource allocation variances, and estimating type uncertainties, what is the actual source of a later task overrun? First, you are left with no low-level knowledge regarding which parameter caused the actual status. Second, and more interesting, padding tasks is shown to support the student syndrome described in the

CC theory and is a well-known personal behavioral phenomenon. This scenario is not well understood in traditional projects. Managing a project using CC estimating logic and disciplined resource management can help improve outcomes without changing the design structure of the waterfall model. These practices are simply a mismatch of project goals linked to how the model is utilized.

Beyond all of the above basic task operational issues, there is one even greater "elephant in the room" high-level issue to understand. This issue spawns from project scope changes. Be reminded that the original predictive model definition assumed that all project scope was known and approved for execution. Scope change management was added as a band-aid to better match reality. However, by definition, any scope change has the potential to negatively effect the approved plan accuracy. If one admits that scope changes are the norm, then that leads to the conclusion that the initial approved plan didn't mean very much as a forecast—i.e., by definition, it does not describe the project completion metrics unless one tracks the status for both original and current values. A process to effectively handle scope changes represents the Achilles heel of the waterfall model. Once scope change has occurred, it is important to recognize that the original plan is no longer valid. Chapter 6 outlined the recommended process, and it needs to be followed if one wants to attempt to properly manage the predictive view. If the project has a formal third-party contracting arrangement, all changes may be added to the contract scope and budget; however, that same formality is often not followed in internal projects. Too often scope changes are buried by simply adding padding to various places in the plan, which completely hides all impact of changes. If there is a need to compare the final plan to the actual one executed, scope values need to be added to the original values. If one thinks about the role of a project plan, is it to show what you hope will occur or is it just a wish list much like putting bars on a Gantt chart and calling that a plan? This line of discussion is meant to emphasize that a model should support an execution process that reflects reality and track status based on the same philosophy. Status views created from false assumptions are worthless and misleading. If the design purpose of a formal project plan is to forecast completion parameters, one can argue from these examples that waterfall plans improperly managed do not do that and do not represent reality.

It is recognized that many of the plan variables are not easy to estimate. Based on this, how much time should be spent creating formal plans that by definition have a significant error? Might it be better to do less planning and set constraint limits to work within, then give the project team and users more freedom to produce the best product possible within defined constraints? This is starting to look like agile reincarnated. Some organizational cultures will resist this line of logic because of the apparent loss of control, but the argument above already described this view as being a myth. Given the design flaws in the traditional formal plan, there is little true implicit control, and the current practice does not help in motivating the team to produce faster.

The goal here is not to tear down the traditional organizational culture but maybe to help recognize what isn't happening and what bad practices are causing. If there was a way to cut down on overall cycle time and produce at least equal deliverables with better customer satisfaction, is that not a worthy goal? The technical answer to this question is obvious. The new integrated model as described here will show a mechanism to improve the current approach, but at the core, the proper approach to this issue is to have an appropriate decision process that deals with issues leading to failure regardless of the management model being followed. The following eight practices will move the classic predictive model closer to the integrated model view:

1. Follow the best practices outlined here
2. Manage resource availability as a key process
3. Use the CC theory as a task-estimating guideline
4. Develop requirements using the MoSCoW concept rather than a singular discrete value
5. Use work chains as sprint-like management focus groupings
6. Status reporting should be more macro-level and probability-based
7. Project goals will be prioritized and customized to the specific project goals
8. Buffers will be used to protect dates; no task padding elsewhere

From this view, the goal would be to have resources in place as needed, track the completion of tasks (or work chains) using Kanban-type progress flowthrough charts, and challenge the team to move forward. The new integrated model will support modifications of this type as a transition strategy to minimize organizational culture resistance.

The points described in this section are meant to create an understanding related to how this model can be used as designed to guide the result toward improved results. When one decides to modify the structure of the model it should be done with an understanding of how that affects the reality assumption. One cannot just say that they are following this model if they do not understand what the model assumes and how it must be managed. There are components of the waterfall model that need to remain visible in the new integrated model.

Agile Analysis

The recognized operational success of agile and its dialects makes this model approach a hard item to criticize. One has to accept the idea that something in this process is causing projects to be viewed as more successful when compared to a waterfall structure. The basic operational techniques associated with the model were covered in Chapter 9, and the concept of loosely defined iterative sprints is now an accepted delivery model. However, this does not mean it will work the same in all project profile environments. The primary difference between agile and waterfall processes is the level of task planning before execution. A key management design question is to decide how much should the scope be predefined before moving into an execution environment. The answer for this is driven by factors such as the type of task involved, the skill of the worker, and possibly the culture of the organization. If the worker assigned to accomplish the task knows what to do from some form of task specification, then deciding to allocate this into the sprint backlog seems to be an appropriate strategy. There is mounting evidence that spending less time pre-defining and having a close relationship with key stakeholders is a proven method to achieve high user satisfaction. However, for a tangible type delivery goal, using multiple iterative sprints to deliver an outcome is not feasible. Traditionally, the process was to decide to either follow the sprint option or the predictive waterfall approach, while at the same time questioning the need for less specification and allowing for more decision-making at the team level. There is now consideration of finding a halfway point solution using some type of looser specification method. It is interesting to review the approach used by Lockheed's Skunk Works in producing high-technology products over the years where the team works out specifics from stated requirements. This is a working example of iteration in a tangible product environment with highly skilled teams, and it does represent a crossover view across theoretical model types. Experiences such as this should have already made a more visible impact on the project industry. More research in identifying the right mix between fixed and lesser definitions is badly needed to operationalize this concept. If one looks at projects where it is not feasible to modify the output then classic agile processes will have to be modified accordingly and that motivation is defined here as "modified Scrum." This is now added to the list of legitimate work execution options. More details on this approach will be described below.

Many physical products do not allow significant variability, so their task definition must be fixed. Nevertheless, elements of the various communication and team-oriented techniques that have been validated in agile can also be utilized in all projects. The agile experience has demonstrated that decreasing up-front specification and having active interplay between the team and stakeholders can produce high usability deliverables. Based on this success, the agile concept has attracted great interest in the project community. As with all good ideas, the key to the expanded use of iteration in a production environment is to find the right method of executing the sprint process. Regardless, a mix of agile principles with a waterfall structure is an attractive view for the future.

Beyond all the agile positives, there are challenges for the approach when applied to a wider array of project types. For example, if organizations require business case-type data for comparative portfolio analysis. Planning processes from the predictive model would seem to be helpful for this role. Second, if senior management demands more deliverable forecast projections for functionality, cost, or schedule, then agile has to add that view to its core process. Another alternative is to recognize that the waterfall planning model is excessive and instead implement some of the agile softer requirement's techniques in its place. The theory here is that the Modified Scrum process would be utilized for the associated work execution. As indicated earlier, this class of work change will have an impact on status-tracking processes. The management level will have to be educated on what this does to the level of status definition available. In any case, a change in status tracking processes is needed across all of the work options as outlined earlier. While the project team level may be quite happy with the iterative tracking approach, that is less true for outsiders who do not understand the less specifically planned deliverable work model.

The main contribution that agile experience brings to the management area is the increased delegation of work management principles related to the sprint process. Concepts such as dedicating resources, short delivery targets, timely communication with a daily standup, and heavy user participation could be of general help in all projects. Assuming the new integrated model allows both predictive and iterative work within the same project, the sprint mechanics will need to be added to the work management process.

Modified Scrum Analysis

Although this modified agile sprint work type has not been extensively tested, there is increasing interest in expanding the use of iterative work units based on the positive experiences using this execution approach. This method will be utilized when it is decided to execute some or all portions of the project using MoSCoW requirements concepts. These task units will be assigned to a sprint queue of tangible deliverables for which the scope parameters are rerquired. In many ways, this is much like a work chain within a traditional model, except the required output is now defined by some level of MoSCoW requirements. In other words, this sprint can be defined as complete if it delivers the M and S components. Anything more is not required. These Scrum units will use known Scrum principles except the required output will have to be delivered in one fixed timebox sprint iteration. An entire project can be defined this way in the new model, although the example here shows this work goal being used for a more mixed hybrid form that utilizes two parallel work queues—traditional and Modified Scrum. This means that the Modified Scrum sprint may overrun its timebox if it does not complete the required deliverables. A previous note outlined the graded requirements technique and that seems to be the prerequisite to making this work. An additional sprint can also be sized to produce residual requirements (more functionality) if that is deemed worthwhile, but recognize that the lower-level output has been previously defined as minimally adequate, and the item could go into production with those specifications. The sizing of these sprints is still meant to be short cycles, but this sprint class may have to be extended in time based on the internal work effort allocated, but short cycles are recognized as an important aspect of productivity. Utilization of the work technique is one of the key success strategies for the new model.

Iteration versus Prediction

One subtle cultural issue observed in agile projects is the difference in the level of management detail. Historically, senior management tended to be somewhat dogmatic in the way they wanted to oversee projects, particularly forecast details. The perceived ability for low levels of control in the waterfall model is one of the probable barriers to change. This model clearly shows task and project-level quantitative values for time and cost. That is a comforting feeling for management and significant time is spent comparing plans versus actual values through the execution cycle. Previous sections have described some of the basic flaws in this belief, but the myth continues. Even with consistent forecast errors, it seems to be more acceptable to look at status in this traditional format. This is significantly different in the iterative model. The fact that the waterfall model fails to accurately predict the actual schedule seems to be often ignored. The current perception of this control view is one of the strange psychological aspects of the modern project and in some ways reflects the immaturity underlying this activity. For this bias reason, project managers will likely continue to be challenged in the future to forecast deliverable outcomes regardless of the underlying model approach. The agile model evaluates results at the end of sprints, and there is less oversight on a sophisticated predefined three-variable success formula. Success is more measured by customer evaluation of sprint results and this can change over time. The reality of a fixed requirement seems somewhat archaic in the face of what we now know about waterfall status processes. The use of a fuzzier scope definition as found in agile would allow more intelligent dynamic work management when the project is drifting away from targets.

The lack of total life cycle planning and status quantification in agile mechanics means that senior management is mostly left to define the stop point for the sprint iterations, which will likely be based on budget, HR resources, or time constraints. Another point to make here is the increased recognition of a more dynamic approach to deliverable success. This may also positively impact the use of iterative techniques as it focuses more on that aspect than the predictive model.

Critical Chain Analysis

Real-world experience with this model is much less than it deserves, but it does have documented success by organizations such as Raytheon where it has been broadly implemented. One can find positive comments in the literature describing improved project cycle times in the range of 20% compared to a traditional model approach. These seem reasonable given CC's disciplined focus on task chain execution. In addition, this execution approach can be effectively utilized for any task grouping that is selected for maximum throughput speed. Another modified usage is to complete selected "pinch" points within a traditional waterfall structure. The major value of this model is in understanding the required process for maximum speed, a lesson that all project managers should know well. The integrated model has modified this idea for use in task subsets of the project called "work chains." An example of this will be shown in the next chapter as the model is described further. One of the resistance points related to CC is its heavy use of buffers and the associated need to heavily modify traditional control logic since the concept of scheduled task completion is defined. The discipline required regarding resource management needed to execute the method is also difficult for most organizations. The sweet spot for this model's use seems most appropriate in subsections of the plan where chains of tasks need to be completed quickly. When viewed as an isolated work chain, the buffer logic is simpler to define and the bulk of the value is derived from improved task speed. In addition to using this in a traditional project model, the general principle may also have similar value in managing an agile sprint. CC concepts offer great insight into effective task management, and for this reason, it represents a useful option for the work management decision.

Infrastructure Analysis

In the ideal case, the project's host organization will have existing formal systems to cover all financial, HR, status-tracking, IT infrastructure, physical facilities, as well as other operational-level system needs of the project. Chapter 11 described a theoretical view of this requirement. For a new project, a review of this support component in the planning phase is important because any missing elements in the host environment will have to be custom supplied by the project with related additional project time and resources. The subtlety of this component lies in any gaps existing in the host organization. Many projects do not recognize the role of this external entity in the project deliverable process. Examples of common organizational support gaps are processes to collect actual resource costs, payroll linkages, procurement, or a change request system. The same is true of centralized computing resources and software needed by the team. A mature organization will ensure that this type of environment is in place. The cost and time advantages of having shared services at the organizational level are significant. Although not mentioned earlier, some organizations are pooling project managers and team resources for better human resource control. These examples illustrate that the project does not stand alone but is often tightly embedded in the host organization. This component will be explicitly shown in the new structure.

Portfolio Analysis

The need for this topic is not controversial, and some semblance of this process exists in a majority of organizations; however, there is a subtle aspect to it that requires understanding. Just because there is a titled function that focuses on project selection does not mean it works properly. Even though strategic planning functions frequently exist, surveys show that they are not consistent in their goal structure and they frequently fail to remain in place. This failure frequency may well be related to organizational entities having decision-making authority over lower-level activities. There is an old saying "I'm from headquarters and here to help." The project level is very territorial and independent thinking. There is extreme bias against the need for help from higher-level entities. It may be all right to provide service to the project but not so acceptable to define whether the project should be approved or not. The integrated model requires this level of control. In order for this function to be acceptable and successful it needs to be perceived as a project service-level activity. Examples of this are assisting with task estimating, project training, or other actions that do not inhibit the project. Unfortunately, the role in the integrated model hits this function at its most conflicting role, which is selecting the best collection of organizational projects within resource constraints. Senior management will have to be actively involved in this regardless of the lower-level resistance. Second, most of the current PMO organizations are departmental in scope. The integrated model requires the scope of this function to be global with the title of EPMO for the enterprise level. This makes the decision process even more complex and requires that departmental-level managers also be major players in the activity with a focus on a global team role and not just focused on supporting their local collection of projects.

Even if the proposed project uses a pure agile development model, it will have to present data to show that it is a better option than other competing traditional predictive projects. In the current vernacular this quantification process is often called a Business Case and the format is well-defined in most organizations. This iterative project proposal must describe both the positive value and negative resource impact on the organization. Also, some indication of risk should be part of this review (i.e., technical, political, resource, etc.). To produce data of this type, some added level of predictive analysis is required. At their discretion, management can ignore these selection rules and just say "get it done" without defined restrictions, but that may be a once-in-a-lifetime event. The accuracy of this planning level is a key to organizational success and because of the heavy potential conflict involved its role will be a challenge. This central project selection function is a vital part of the overall project success equation as a bad selection generally means wasted resources.

The author once asked a PMO manager in a large organization how many project proposals they dealt with. His response was 4000! It is clear in situations of this size that project approval has to be organized in some formal process to escape the chaos of project proliferation. Recognizing that this next statement is hard to verify, some consultant organizations will say that a *laissez-faire* approach to project selection can result in a waste of 20% of the total expenditure. As an example, many years ago an organization installed a new email system unknown to outside departments. Later, other departments did the same and within two years there were nine such isolated systems in place that could not talk to each other. The resolution of this required a major project remediation effort to homogenize this collection into one standard system. Many turf wars were involved in deciding which system would be the standard. This situation could have been avoided with proper portfolio management and project control.

Here is the key memory example for the portfolio area. *Doing the wrong project correctly is still the wrong project.* The key requirement to make for this management component is for project approval decisions to be formally controlled by management at the portfolio level and then overseen as the project cascades down into the lower-level decision processes. Recognize that the overall proposed and active project population operates out of a single constrained resource pool. For this reason, a runaway failing project at the execution level has an impact on the ability to approve other proposed projects. So, macro-level project status data must include both the active projects and the proposed ones. This oversight role will require the active support of senior management to work, but it needs to be recognized as a necessary organizational success process. The goal of this component of project management is to optimize organizational goals by selecting the best collection of projects within constrained resources. Organizations that do not formally recognize this level of project activity are not managing but reacting to disjointed requirements.

Unfortunately, this segment of project decision-making is one of the most error-prone and conflict-driven of all management segments described here. First, the ability to predict the value of a specific proposal is difficult, and accurately estimating the resource implication of that idea is equally difficult. Nevertheless, this is the requirement and yet another example where management and key stakeholders' active participation is needed.

Multiple Work Delivery Queues

The concept of managing variable work definition is not a typical work management approach since traditional models do not readily recognize this ability. The construction case study example previously described in the text provides a visible example to show how multiple work delivery options could improve customer satisfaction. The key point here is to highlight that various tasks can have variable profile constraints and goals that can be best matched by using different delivery options within a single project structure.

Specific points or recommendations made in this chapter are based on the various delivery model characteristics that were outlined in the previous text sections. Some of these points will be controversial and the reader will have to evaluate these for their specific environment. These points are driven by the author's personal experience. The majority of the new work management views introduced here have been created from the observed mismatch of project goals within the classic models. None of these models fit the global breadth of project types, breadth of management, multiple work options, or delivery goals.

Figure 13.1 shows the four options for executing a work unit (or work chain).

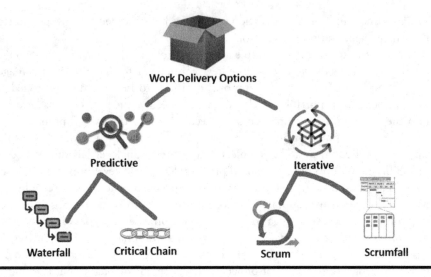

Figure 13.1   Delivery work queue structure

Predictive Delivery Queue

Work that is moved to the predictive queue (reasonable requirements definition) needs to be evaluated further for delivery goals. If the goal for this group is time compression and resources are available, the work units will be earmarked for Critical Chain logic. This can be applied either for a linked chain of work or a collection of tasks within the WBS.

If the CC option is not appropriate, the tasks will follow the traditional waterfall model rules. However, there is still some flexibility in this task sequence regarding the level of scope definition rigor before execution. The resulting view can be a pure waterfall process or a modified Scrum sprint logic using MoSCoW requirements (described earlier). The exact execution process is dictated by the degree of scope formality. Task durations can be either fixed or variable based on the estimating logic chosen. This project management approach can be executed as lean or with full formal traditional bells and whistles. The question of not using a formal scope definition may be against local policy rules, but this is the time to start dealing with the fact that this does not necessarily add accuracy to the final result. Suppose that some segment of the project had significant slack and the associated resources were very mature. In this segment, the correct decision is to require less status oversight and let the team manage the work requirements within defined constraints. The concept of flexibility should now start echoing throughout this structure.

The time compression segment deserves a few comments beyond just following the CC model. There are many situations where time is a major deliverable goal, but resources are not available to the degree required for full CC treatment. In other words, this can be a strategy within portions of the overall structure. Think of this segment as a sub-chain of tasks for time minimization focus. In this segment the management strategy is to execute the following mechanics:

- A resource priority focus, even if not to the rigor of CC rules (estimating and resources)
- Time estimates set at 50/50 probability
- Disciplined workers focus on active tasks (no multitasking)
- Appropriate buffers inserted in the plan to cover overruns

Recognition of a full CC model or the subset chain view is based on the availability of resources and the priorities of the project. At this stage, we have background knowledge of the expected results from employing either of these options and that is the essence of the new model—i.e., different results from different management decisions. This scenario reflects a significantly different work environment from the traditional waterfall. Note that the various work management options are not defined in the model but based on the options dictated by the project.

Iterative Delivery Queue

The iterative work queue is handled by moving selected work units into an iterative execution structure. The two model choices described are basic agile and a modified Scrum (see Figure 13.1). In both cases, this work queue is managed using sprint work principles. That means a fixed-time sprint will be used to execute the classic agile sprint iterations and a modified sprint will be used for the predictive units with MoSCoW scope definition. The modified predictive work queue is designed to take advantage of the proven iterative success model in a slightly modified way. In this scenario, a modified Scrum sprint must achieve all of the base requirements before it can terminate. Once an organization has learned how to utilize this approach, it could well become the dominant work strategy. But we still believe that many project types still need to be managed as a traditional scope model, supported by the use of best practices to minimize the reality gaps.

Summary

The individual points made in this preliminary model component overview have summarized key global management components that are recognized in designing the integrated model. The following list is a summary of the resulting new model characteristics:

1. All projects do not have the same work profiles and therefore defining the characteristics and delivery goals is a part of the model decision process.

2. Each project may have multiple task characteristics that can best be executed using the four different work execution options.
3. Selected CC task management characteristics can be applied in various segments of a project to improve throughput.
4. There must be a feedback status tracking process link between the active project set and the proposed portfolio since these two layers compete for the same constrained resource pool.
5. An organization's project support infrastructure is an important part of the overall management view. The ability to link into shared services from the host organization minimizes internal project resource allocations.

This chapter has been a capsule overview summarizing the key model components that are needed to describe the integrated model. At this point, we have a reasonable overview of the target ideas that need to be packaged into a management structure. In addition to the model structure, various failure-related items also require management focus within the core decision structure. Project tasks can be defined using a mix of multiple execution techniques, and these can be flexibly embedded in the structure. The process to accomplish this involves matching the project characteristics and goals to each work unit, rather than forcing the defined tasks to fit some preselected model. No one model fits all projects, and the management mantra must be to map the target project using appropriate work execution options,

Reference

Richardson, G. L. and B. M. Jackson. 2019. *Project Management Theory and Practice*. 3rd ed. Boca Raton, Fl: CRC Press.

Chapter 14

The Integrated Delivery Model

Introduction

Previous chapters have reviewed various processes related to producing successful project results. One of the main focus areas is an of how to utilize validated classic models using a modified execution strategy. Key processes were analyzed and dissected for usable components. The previous chapter descriptions should have satisfied the notion that all projects are different and a modified flexible model is required to match this variability based on a specified project profile. Following the original design specifications, a supporting set of core components was described in the previous chapter. Each of the identified components serves an important role in managing the evolution of the project from the initial approval decision to the final delivery. Based on the variables defined for the specific project, it is necessary to customize the management approach for each project. One key observation from the model development process is that the classic model's fixed management structure does not fit reality, so a customization work process is needed. Also, the resulting model will not be a cookbook of fixed steps for the same reason. From data outlined in the previous background chapters, it is concluded that all of the current management models were static in their view of task work execution, and this does not fit the profile variability observed in projects. This led to a design strategy of matching multiple work execution options within the same structure and being driven by the project goals. This chapter uses the results of this research to begin formulating a model that fits the proposed design. This new structure will morph itself around the project's profile with the goal being to produce the defined deliverable and avoid band-aid gap patches that distort the integrity of the model. The resulting structure utilizes selected pieces of validated classic concepts and work processes but will use them to match the project profile. The term Lego block is used to logically describe how these pieces will be "snapped" in place as needed to satisfy the profile goal.

Design Features

Thirteen design principles for the integrated model are summarized below. Most of the items on this list can be considered existing best practices, except for two processes related to project profiling (#3) and task management (#7):

1. Project proposals are comprised of technical, financial, and other relevant data.
2. Investment decisions are based on an evaluation of the project proposals against a set of criteria, such as alignment with strategy, costs, resources, benefits, and risk.
3. The defined project profile is mapped to a proper management approach.
4. Manage the project using known best practices.
5. Macro work units are designed to match the defined goals.
6. Scope and risk reserves are used per defined best practices.
7. Manage task execution decisions based on defined ranked delivery goals.
8. Focus on generating delivery functionality as early as reasonable.
9. Modify status reporting to focus more on customer satisfaction and less on fixed calendar dates.
10. Utilize buffers to protect overruns and planned completion values will be described by range values.
11. Recognize opportunities to decrease the formal upfront planning level in return for more team delegation and improved morale.
12. Train project teams to utilize the new model.
13. Formally close the project and archive files for future shared use.

Decision Block Architecture

The skeleton of the new model is represented by multiple layers of linked decisions starting at project approval and ending with closing the project. Even though six decision packages are defined in the figure, more decision blocks can be added if necessary to fit the project needs. This is stated to highlight the notion that a project may have an unusual need, and such decisions need to be represented in the plan structure. An example, a large post-delivery deployment step is significant enough to justify adding this process to the overall view. In addition, two success-oriented support blocks are added to the core decision

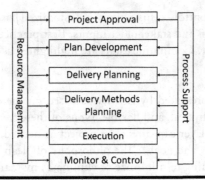

Figure 14.1   Integrated model block diagram

blocks. Project decisions flow downward through these blocks into a flexible execution layer, and the two supporting components represent ongoing global management functions. A block diagram illustrating the decision logic skeleton is shown in Figure 14.1.

The underlying logic of the block diagram is important to understand as it represents the decision flow through the life cycle. A brief description of each block role follows:

1. Project Approval—This is the strategic decision layer that controls the portfolio segment through the EPMO function described in Chapter 12.
2. Plan Development—This decision layer takes the project's approval data and begins the process of defining how the project will be structured. Various preparation processes profile quantification, macrostructures, risk decisions, and general management guidance are outlined here.
3. Delivery Planning—This is a logical extension of the previous step. Expanded details for the active portion of the project are produced. This would include expanded planning for requirements definition, schedule, budget planning, resources, and organizational support. Specifications on status reporting are defined.
4. Delivery Methods Planning—The project team is now in place and begins work on internal training and developing strategies for delivering various project components. Details regarding the use of specific execution methods are resolved. Work plans are finalized. Testing, customer acceptance, and project close activities are recognized. Milestone events, delivery constraints, and final plans are reviewed with management. Decisions on work execution options are made.
5. Execution—Work is executed per the plan. Best practices are used in the work process. Processes related to risk management and status tracking are implemented as defined.

6. Monitor and Control—This activity is customized based on the original control specification and stakeholder inputs. Best practices are used here as outlined in the text. Communications are highlighted as a major project failure process.

The decision flow represented by this list is based on the evolving nature of the process—logical to physical, and flexible. None of these steps are frozen regarding the outcome as each layer can direct the project as needed, guided by the project profile (characteristics and goals). The philosophy of "lean" is implied within the decision blocks, meaning to minimize or eliminate any management activities not judged necessary. This includes topics such as risk analysis, status tracking, documentation, and scope definition rigor. As decisions flow downward, the executing level focuses on producing the defined output using options selected from one of the four defined methods. The execution method chosen will be matched with the proper output goal as described earlier in the text.

It is valid to think of this model as a roadmap to aid in making appropriate decisions based on a flexible goal set. The block diagram represents the high-level view of the layered decisions. Underneath this view are two more philosophical aspects that need to be understood. First, the decisions shown are meant to be made with a specific project profile in mind and not a static set of steps and detail level. At initiation, there is no preconceived notion as to how best to execute the effort. The terms "lean" and "flexible" best describe the decision block philosophy. Second, there is a host of known failure-related best practices that have been outlined in the text and others described in the industry that collectively needs to be reviewed within these decision blocks. The text has provided examples to show why these are typical sources of failure. It is assumed that proper training will be supplied to the team to support these goals and thereby avoid repeating the same negative factors as the industry surveys indicate to be common. Eliminating repeat negative practices needs to be a management focus area.

The main philosophical difference between the new model and those examined in previous chapters is it has a design structure focused on the target project's characteristics and then uses that profile to customize the life cycle and select different work execution strategies that best support the project delivery priorities. This statement best represents the definition of the new model.

The potential increased productivity from using the sprint model principles is recognized, but other models are also shown to be appropriate in different circumstances. The approach of mapping the project's execution process around the target project's profile goal essentially inverts this decision process compared to the classic static view found in all others reviewed. Handling variable project characteristics in the classic models seems to be done by patching various processes on top of the base model, which essentially compromises its design structure. Patches in the integrated model simply represent an explicit decision to add that process (or leave it out if it isn't needed). In addition to this, even in the case where some process is added, it is only added to the degree necessary. Examples of this are the level of risk assessment, types of documentation, level of status reporting, and most important the rigor of scope definition that can lead to a different execution option.

There are other subtle decisions required in the layers. For instance, if there are aspects of the project environment that can affect success, they need to be visible in the management structure. As examples of this, none of the classic models show that a project selection is a key success-related event, that resource management directly affects the output, or that the host organization is a success factor. Admittedly, traditional theory sources describe the portfolio management role and also describe the role of resource management, but none mechanically link that to the working model. All three of these sources directly affect success and for that reason need to be part of the visible management activity. All of these factors need to be shaped around the target project needs and none are viewed as fixed. Failure to recognize and deal with these facts can lead to project failure.

Classic management theory outlines various activities or "knowledge" areas that need to be followed in producing successful outcomes, but the flaw in that logic is that not all projects need the same degree of these, or maybe none at all. Some projects have significant risks that justify both a formal initial assessment and an ongoing tracking and control process. Other projects with familiar backgrounds would not need much resource expenditure related to this topic. The logic block related to this question triggers the decision, which then flows downstream for further action as needed. This represents the flexible design approach which may be confusing to the traditional manager but is an important aspect of the new design. Too often, the rigor of classic models lends to non-productive extra work which the technical staff knows is not needed. Another subtlety is the increased level of trust implied by moving more decisions to the working level and making them more directly responsible for the outcomes. These examples show why the traditional model structures are not well aligned to managing a custom deliverable goal set, or flexibly performing the right processes along that path. This form of project decision-making is meant to move decisions to the best level for execution.

There are three process areas where the new model has a different management philosophy. First, in the risk area, the model allows complete flexibility regarding the level of risk assessment to be specified. The intent of this is to stay as lean as reasonable, and then assign formal risk owners in selected areas within the structure to be key life cycle monitors overseeing that area. Second, the negative impact of padded task estimating has been frequently shown to be a negative factor, so tasks will be estimated using the Critical Chain approach, which then necessitates the use of buffers for anticipated overflows. Third, requirements will be more focused on the multi-grade MoSCoW approach described earlier. This approach will support the use of modified execution options such as Modified Scrum predictive sprints. Finally, all of these modified management approaches will have an impact on the status reporting process. The fixed task schedule approach found in the waterfall model is no longer valid (and it never really was). All status reporting will now be modified to either range or probability formats, and the evaluation of buffer status will increase in usage for status analysis.

Even though this new model description may be troubling to some based on its flexibility and multiple work queues, this approach is necessary to map the proper delivery approach to the project. Previous descriptions have highlighted why each of these more flexible approaches supports the design objectives. Loosening delivery into grades has the potential for opening up more work flexibility and higher productivity. The logic to support this approach is based on experience showing that the traditional task specification process is time-consuming and often contains a significant inaccuracy. The biggest improvement in risk exposure comes from creating a risk culture in the organization and assigning knowledgeable employees the responsibility for this, rather than using extensive formal risk modeling. When a project is evaluated as having a high-risk potential, then the full risk analysis and control mechanisms should be pursued.

One other component not well reflected in the traditional model is the view of resources. One gets the impression that they are somehow just available. Nothing could be further from the truth. Appropriate access to resources only happens from significant management attention.

When the goal is to compress the schedule in a particular part of the project, the Critical Chain theory description outlines the focus and discipline required to execute that idea. Even in the more traditional view, not having appropriate resources in place causes a schedule overrun, even with a perfect plan. Once the conversion is made from padded time estimates to the 50/50 model, the pressure on human resources will be even higher. This new approach will mean that there is no longer a target date for task completion, but more like everything going as fast as it can. This is a quite different project environment than found in the traditional predictive one. The new estimating approach will result in more schedule overruns and protective buffers. As projects move into this more turbo mode, it is not hard to envision some of the new team management issues that could follow. There are other management aspects related to the resource pool, but this should suffice to justify it being recognized as part of the overall block view.

As a final point on the management environment required with the new model, project failure can be created in a myriad of areas and this list is very long. Project managers in this new environment must be well aware of the failure root cause factors and monitor those areas carefully. Comments related to best practices have been scattered throughout the text, but the summarized checklist below of these key items is a reminder that the model recognizes such processes as important to be followed in the various decision blocks. These key best practices are to be followed regardless of the underlying management structure. This reminder list is shown here to emphasize this aspect of the process and the value of a learning organization concept. This may well be the reason these items keep recurring in the industry surveys. A learning organization should aspire to not repeat negative events and seek out the root cause of such.

Project success has many similarities to sports teams that are successful. Winning teams have a visible competitive culture and a superior confidence level that they are better than the competition. This same attitude applies to project teams and often means that they are willing to try new approaches as described in this text. Being the best is a short-term view if one does not pursue improvements.

Hopefully, the light-hearted metaphor examples described earlier help with the logic outlined here. At this point, are still trying to define the whole elephant and the things that will make him healthy. Hopefully, the two previous seemingly unrelated stories make more sense now. Here is the current translation:

1. The drunk looking for his keys under the streetlight when he lost them somewhere else (in the dark)— this is analogous as looking for a new management model based on the project characteristics rather than a fixed structure.
2. The six blind men try to identify the elephant by only looking at parts— seeking an integrated approach.—an integrated view rather than isolated segments

The new model has expanded the work view to encapsulate the whole process and it has looked for the keys in a different place. Models such as the one described here are meant to help one sort through the complexities by looking at a large (project) elephant and not knowing which piece to start with. We are now sensitized to the idea that this process is difficult to sort through and thus far no one has been able to draft a fixed set of decision steps that lead to a high level of delivery success. There is even confusion now as to how to measure success and the evolving definition in the new model recognizes that complexity as well. The same can be said for work execution strategy. Think about these sample driving questions. Is the goal of this project to finish quickly? That doesn't fit the waterfall view very well. Is the goal to manage a large high-technology project? That doesn't fit agile very well. The common view across all of the classic models examined is that they do niche things reasonably well, but they do not do all variations well. In addition to this, many projects have characteristics that fit multiple work execution options that are ignored since the model being followed does not support that. The integrated model opens up that needed view. There are many more examples where different project characteristics need to be matched, and the structure outlined here offers that. With this as a somewhat philosophical background, let's look deeper into the decision block structure.

Model Components

During the life cycle of a project, various decisions steps need to be made to guide the future direction. This has been characterized as moving from a logical view of functionality to a physical work view that produces that functionality. In the middle of this process are numerous decisions that need to be made that affect the future direction and resulting outcomes. The block decision structure represents the core skeleton of the life cycle that can be mapped to all project types and expanded to fit unique situations such as a major deployment requirement. The integrated management model has compartmentalized this decision set into six management groups. These are:

1. Project approval—a process that deals with organizational vision and strategy, plus a process for managing the selection of project targets.
2. Plan development—a mid-level function to evaluate how best to package an approved target (i.e., phases, programs, timing, etc.).
3. Delivery packaging—development of the high-level WBS which includes an elaboration of requirements. The project profile is documented along with delivery goal priorities.
4. Delivery methods planning—Based on factors derived in the previous step, this process involves more details views of the delivery format. This is the core process that identifies the proper way to execute the task defined by the planning process.
5. Execution—Selected work units are completed according to the delivery plan. This can result in multiple work streams. Within the same project.
6. Monitor and control—Execution results are tracked according to the control needs established for the project.

Support Components

Two support entities are attached to the six core model decision layers. The organizational support function role was described in Chapter 11 along with the reason to be recognized. The second support area involves activities related to resource management. From a supply view, this resource is the technical key to driving the model process. Traditional predictive models describe a function called capacity management, but most of these project types do not rigorously follow that theory in practice, which may well be caused by the resource supply dynamics related to scope change and task estimating processes. Beyond the raw issue of quantity and quality of resources needed for the project, there are also many other human management roles vital to success. As stated earlier, this text is primarily focused on the timely resource supply issue and not on the management techniques to support a high-performance team. Also, recognize that there are many alternative project organizational forms for this function, but the focus here is to supply the various process layers with proper human and other tangible technical support.

Core Tools

Within the project decision structure, various schematic and graphical tools have long been used to support various aspects of the management process. There is no intent here to redefine these, but rather to use them as needed based on the process requirements. Some of the tools fall into the category of describing high-level decisions of the project, such as showing how the project might be broken into phases or subsystems. There are numerous tools to show work sequence schedules or the status of work. The new model does not change what these models can help describe, rather it increases the flexibility of their use.

One of the controversial tools that the professional audience has for years disparaged the lack of sophistication is the early 1900s classic Gantt chart. The technical merits of this chart are indeed suspect, but the communication value and external stakeholder acceptance of it represent a key lesson. This is the view that the external audience wants for the project, so it will be a challenge to change that view. This format is easy to read and will stay as a preferred communication medium, so it is up to the internal project team to ensure its integrity. In many ways, the Gantt view becomes the artifact showing the schedule window for the project that all can understand. However, recognize that the flexibility of task estimating recommended in the new model does not easily translate into a fixed completion date. If these practices are utilized, the question becomes how to represent schedules in this environment. The preferred project status communication format would now be to change completion status views to probability displays as shown in the text. In fact, given task variability, a fixed date forecast does not fit any of the models. The question here is how deep into the project's internal details the external audience needs to be. In this regard, the Gantt chart is a legacy format that should be minimized in favor of the new probabilistic status approach. Examples shown in the text have illustrated the role of formal milestones protected by buffers. This is a compromise view of the traditional status display.

WBS

The tool that fits the new design best of all is the WBS. This format is familiar, flexible, and understandable, for representing various plan views. These hierarchical diagrams have long been used to show project structures such as programs, phases, subsystems, and tasks. The flexibility of this tool makes it an easy choice to keep in the toolset. There is no intent to standardize the role of the tool, but sample uses of it will be illustrated later. It should be looked at as a general function flexible partitioning schematic diagram. At the higher block decision levels, the key box groups could be used to outline the overall project. One typical grouping at this level is to outline project phases. Later elaboration decisions might open a phase-level view to show various work initiatives such as requirements definition, risk assessment, or other planning-level activities. The lower levels in the structure can then show the execution strategies for various work units. At this level, the structure begins to resemble the classic view of a WBS. It is also recommended to associate these decisions with a formal data store. In the traditional model, this is called a WBS Dictionary and the role here is similar.

A sample WBS structure is shown here to illustrate the concept of work packaging that occurs at the mid-level of the elaboration process. In this example, the previous decisions related to project phases can be seen and multiple functional groups are defined around their delivery strategy or management focus. Also, there is a project goal characteristic attached to each layer. At the lower layer, it might be to show time compression or iteration goals. Note the further down the decision hierarchy, the more physically work-oriented the decisions become. Logic related to this structure can be attached to the WBS Dictionary for examination and/or approval. For example, the schedule plan for the phases might be challenged and changed after further examination of the work specifics. The WBS schematic in Figure 14.2 shows the project's high-level work grouping.

- Three major phases (1.1, 1.1.5, 1.2, and 1.3)
- Two major work groups identified for phase 1
- Two major future phases are planned (1.2 and 1.3)
- Two detailed and related requirements groups are outlined for 1.2 and the related 1.5)

The meaning of a project "phase" can be another separate project, or in this case simply major subsections of the existing project. Carrying the elaboration example further, assume that additional details are defined such as duration and work sequence. Figure 14.3 shows a flattened table WBS with additional work specification details.

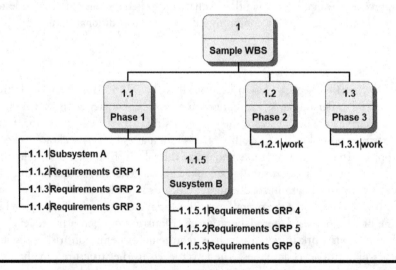

Figure 14.2   WBS showing high-level packaging decisions

WBS	Task Name	Duration
1	**Project Summary**	
1.1	**Design Phase**	
1.1.1	**First Design Phase**	**10 days**
1.1.1.1	Start Milestone	
1.1.1.2	**Design Task 1**	**10 days**
1.1.2	Second Design Phase	33 days
1.1.2.1	**Design Task 2**	**20 days**
1.1.2.2	Design Task 3	15 days
1.1.2.3	Design Task 4	5 days
1.1.2.4	Design Buffer	3 days
1.1.2.4	Design Milestone	
1.2	**Execution Phase**	**29 days**
1.2.1	Task 1	20 days
1.2.2	Task 2	10 days
1.2.3	**Milestone**	
1.3	**Testing Phase**	**17 days**
1.3.1	Test Task 1	10 days
1.3.2	Test Task 2	5 days
1.3.3	Cust Acceptance	10 days
1.3.4	Project Buffer	5 days
1.3.4	End Milestone	0 days

Figure 14.3　Elaborated work process

In this view, the requirements for each phase are now expanded from the schematic WBS view. Notice the expanded definition of key boxes (bolded WBS codes):

- First design phase
- 1.1.1.2 Design task 1
- 1.1.2.1 Design task 2
- 1.2 Execution phase
- 1.2.3 Milestone (major checkpoint)
- 1.3 Testing phase

The concept here is to use the WBS as a macro-level work view guide, and then iteratively add details to that view as more is defined. This is the elaboration idea! The major project segments have meaning to the project or management groups, so they are highlighted here. Each of these is considered a major work delivery group that may have a defined execution strategy specification for status reporting, risk, or some other factor.

Figure 14.2 also illustrates how the WBS structure helps to outline the major details of the project. As the elaboration process continues, the data detail increases so the view collapses into a more table-oriented format. It is important to think of the WBS in both schematic and data formats as both have significant value in the process.

This WBS schematic format focuses on the phase and requirements packaging decisions, while the more data-oriented view in Figure 14.3 contains execution-level details that can be used to produce an initial plan that can be translated into a Gantt chart view as illustrated in Chapter 8. In this case, the resulting Gantt view would have schedule integrity since it is based on planning specification data (i.e., duration and task linkage). Buffers and other control items would need to be attached to this view. These two planning artifacts illustrate again how one level decision leads to another linked one—phase structure to the initial project plan view in this case.

Moving on with the elaboration process, the next phase of the planning process will be one of examining the work required for each of the line items shown. Note in Figure 14.3 that task duration estimates have been added to illustrate incremental decisions. We are assuming here that these are 50/50 estimates per the Critical Chain model. Also, the execution method for each task is recorded. Finally, a sequence code is attached to each task to indicate the order of execution. At this point, we come to a very subtle point. Many times, the schedule produced for a project has no internal integrity, meaning that no consideration was given to items such as sequencing, resources, realistic estimating, or a host of other factors. In this case, the elaboration process is linking the high-level view to the lower-level task execution level. This project's work tasks are now shown in a defined sequence derived from the higher-level view, so there is internal integrity from that exercise. This resulted in a deterministic schedule that will now have to be elaborated to match the execution overrun anticipated. Ignoring this point for a moment it is useful to first produce a traditional-looking schedule for general sizing purposes since computer software can do using the following four parameters:

1. WBS reference code
2. Title
3. Duration estimate
4. Sequence code (predecessor)

Figure 14.4 shows a sample output of this process.

This is the essence of the predictive model with fixed duration estimates, so we know some of the inaccuracies that this view represents, but it is an interesting starting point for the project if enhanced with appropriate buffers and sprint recognition. The calendar dates indicated here are not going to be valid according to the 50/50 estimating rule. Once buffers are inserted, the plan can be used as a high-level roadmap for the project but is not appropriate for status tracking at the task level.

When iterative work execution options are added to this view, the concept of status reporting changes for the sprint units to other status tracking methods may fit better. A popular core tool for iterative status tracking and even some predictive areas is the Kanban chart. Figure 14.5 shows a simple version of a Kanban chart.

Other work packaging decisions may require additional changes to this format, but this represents a core starting place. Views of this type can also be modified to show periods where high-risk activities are planned, budget details, or resource capacity issues. It is also possible to generate probability distributions for schedule and budget completion forecasts using computer software. Samples of this format were described previously.

One more point can be made using the work plan shown in Figure 14.4. Suppose the delivery item defined in WBS 1.1.3.2 (Design tsk 2) was to be executed using the Modified Scrum execution option. In this case, the schedule bar linked

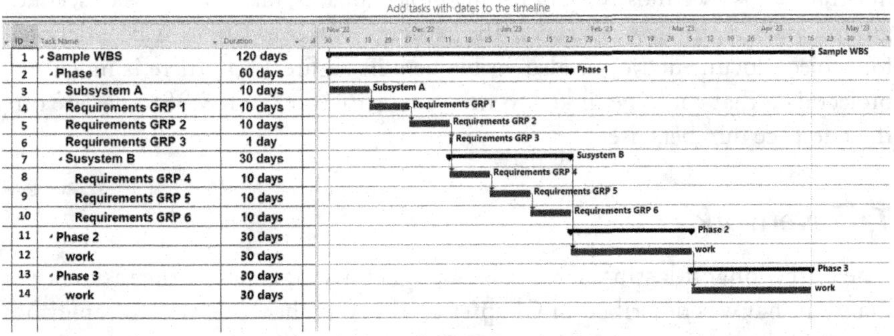

Figure 14.4   Converted high-level project work plan (from Figure 14.3 data)

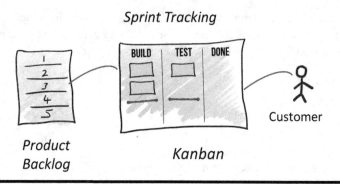

Figure 14.5   Simple Kanban chart

to that work unit would be changed to show that it was a different work option and it would be managed as a Modified Scrum sprint. This decision might cause a change in duration shown and would signal the new execution method. The work plan would be adjusted to fit those decisions. This work execution decision also signals how the delivery requirements will be managed (MoSCoW based), Review the rules described earlier for Modified Scrum as this is an important concept in the new model. The associated task (Gantt) bar now represents a sprint rather than a traditional work unit for execution. Previous chapter descriptions illustrated how this dual work option is handled mechanically.

As the project continues to unfold (elaborate), the decision level becomes more specific regarding work execution mechanics and options. For the predictive segment of a project, the WBS format would migrate toward traditional task-level lineages, while in iterative portions moving toward using a sprint backlog to define work sprints for the chosen items. Scheduling of these will be done in parallel with the predictive tasks, but only the required delivery completion dates would be relevant across these two groups.

In the integrated model, the WBS format is viewed more as a general-purpose decision grouping and work packaging communication tool. As this example has illustrated, it is used mostly for outlining project phases, major subdivisions, control groups, and eventually work tasks. The existence of a companion WBS Dictionary for project documentation is also visible here in Figure 14.3. There is much more project data that could be linked to this same structure, and WBS codes are often the reference numbers used.

Task Network

One of the longest-lasting core tools in project management is the task network diagram that was described in Chapter 8. This was shown in the example above as providing a task overview of the project and a sophisticated technique to calculate a schedule based on time estimates. This is the closest method available to produce a macro view of the project. Microsoft Project and other similar software programs are based on this task linkage design, and this approach has wide use in the industry (with all of the associated gap errors indicated in the text).

Kanban Charts

Kanban's roots originated in manufacturing status tracking where throughput tracking is a primary goal. A sprint is activated from the product backlog and its status is shown on the Kanban. More tracking steps can be defined, but envision this as a visual method to see the basic status of the sprint. A Kanban can also be used to show the status of any type of work task collection or work chain.

Other Tools

Beyond the core tools summarized above other specialized tools can be brought into play at the discretion of the project manager. For example, if a full risk assessment is selected, this brings with it added documentation and status. The use of a risk register becomes a standard artifact, along with other documents prescribed for this process. There are automated tools to track risk event schedules and various recovery documents for the critical process. Recognize that this process is the most immature in modern project management and is often marginally pursued or documented. The suggestion for minimal oversight for this area is to identify and record major risk areas and assign project members as risk owners for those areas. Also, the decision regarding how formal to make the risk assessment is a critical one in the early planning phase.

The role of formal documentation is a common point of conflict. The selection of formal documentation artifacts to be produced should be covered in the Delivery Planning phase. Beyond that, the project team needs to evaluate any additional documentation requirements related to the team, management, or stakeholders. The project profile should guide this decision. Most technical professionals do not relish producing such material, and this decision question is similar to how detailed the planning documentation should be. One technique that can help with all documentation is to use an extensive template library. Also, modern text editing software along with standard templates can produce adequate technical documentation with much less effort. Content support for this can come from a project digital library from which both standard templates and project historical artifacts can be retrieved. Learning organizations gain value in their ability to retrieve archival data for reuse. The value gained from this type of environment cannot be overstated.

Each of the tools and items described here should be evaluated for each project and used as appropriate for the target project.

Decision Layer Roles

The architectural block diagram has outlined the aggregate flow of decisions through the life cycle. This section will serve as a reminder of the required attributes of each layer and how this element links to the success factors described.

Strategic Visioning

It has been increasingly recognized that senior management of the organization has a responsibility to ensure that initiatives undertaken by the organization are appropriate from a resource expenditure and competitive strategy viewpoint. PMI defines this decision layer as:

> *a structure that standardizes the project-related governance processes and facilitates the sharing of resources, methodologies, tools, and techniques.*

(PMI, 2017, p. 216)

As indicated in Chapter 12, the role of projects in organizations has become a major strategy for getting things done. Projects can spawn like weeds in a garden if not controlled. This point seems to be somewhat well recognized in industry today as a majority of organizations have a named function for this role. However, looking more closely it seems that the function is less than effective because of conflicts with lower levels that resist the centralization of this decision. There are three characteristics of this function necessary for success:

- It has to be actively managed by senior management.
- It needs to be for the entire organization since projects cross departmental boundaries and consume global resources (i.e., EPMO).
- There must be a linkage between the high-level decision process and the status of the active project so the allocated resources can be moved between these two levels as needed.

There are two distinct roles for this level. First, the analysis of proposed projects is under the purview of the plan development group, which supports the process of quantifying proposals and building Business Cases for each that pass through a preliminary filter. Second, the actual approval decision is the responsibility of senior management. Specification of overall roles for this function is more difficult to assign than theory would suggest. On the surface, this function should be the "king" deciding on all project events, but that does not seem to be workable. It should be recognized that the centralization of functions is often found to be less effective and that is the case here. A support function for the project approval activity is titled Project Portfolio Management (PPM). The operative challenge for this management layer is to assist in creating formal organizational goals. Without this, the subsequent decision process has no supporting decision data to follow.

The next decision conundrum is to imagine hundreds of proposals that have been quantified by business value, time, budget, resources, and risk levels. Ideally, the decision process should take the highest-value project and work down the list until the resource constraints are reached. That is a valid theory, but the accuracy of these data makes that more of a challenge. Also, the tactical versus the strategic scope of projects adds to the decision complexity. Finally, the RGT phenomenon outlined in the PMO chapter further complicates this decision. A true return on Investment calculation for a Transformation option would only be evaluated years later after the fact. For purposes here, the role of an EPMO function is vital to organizational success. One needs to look at the industry failure rates for this function to see that there is more involved than the simple logic to say it is needed. It is up to the organizational leadership to understand the value of the function. There will be resistance to it from the lower levels who view it as an inhibitor to what they wish to do.

There are notable examples of strategic visioning organizations that carried on a project-type venture for years with no clear return on investment, with the belief that some initiative would achieve a completive advantage. As an example, the Toyota Prius project had been underway for several years waiting on the price of fuel to reach the point where it could be brought successfully to production where it captured most of the electric vehicle auto market. Amazon and Apple's market success experiences illustrate the value of this class of decisions. A key statement to close out this segment of the model recommends that organizations must have a workable process in place for project selection and control if they are to be successful with this activity. There are many structure options available to add to this core function, but they should be carefully introduced to avoid an excessive negative reaction from the lower levels. The main deliverables from the strategic level are to select and approve the correct project initiatives and to act on any failing efforts underway in the execution layer. This second status-tracking role is important to save resources for newer initiatives.

Delivery Planning

The input to this decision layer is an approved project initiative from the EPMO function. Earlier plan variables now need to be validated and any changes reported back to the EPMO and PPM functions. Beyond this, the major issues to deal with at this point are project profiling, risk, and aggregate packaging of the effort. In some cases, a major delivery structure needs to be grouped into some logical form for execution. Another planning aspect is the resource side of the equation. Internal components such as hardware, software, networks, major subsystems, or even the level of requirements grouping. This decision moves the decisions further down the road for deciding what types of work execution best fit the requirement. This is the level at which the WBS defines the work execution decisions based on lower levels of value analysis and/or work focus.

Each of the delivery packages identified at this level should have decomposed data regarding value, schedule, budget, risk, and resource requirements. All of this data will be updated in the organizational project database. At this stage, the project is not yet ready for execution but is at the Work-In-Progress (WIP) stage. Beyond the high-level packaging activity, one other decision at this level is an indicator of priority, which is also part of the packaging logic as well.

Here is a hypothetical example to provide some reality view for this decision layer. Suppose some competitive target had been identified and the EPMO wanted to have some portion of it into production as soon as possible. Completing the production version is estimated to take two years. However, by producing what might be described as a production prototype with only essential requirements, that version could be produced in six months. So, the packaging decision is to build the prototype version, then produce the production version with still limited requirements, and in the third phase produce the full package. Packaging decisions such as this are made to recognize both the technical grouping and the business delivery aspects. One can see in this case, the focus is on getting the minimum required functionality into production as soon as possible. Too often a decision of this type will be made only after completion of the full requirements, thereby losing the short-term business value or possibly even missing the competitive window completely. This level of decision-making is second in importance only to identifying the correct targets.

Delivery Methods Planning

The role of this layer is to evaluate the driving factors that most influence the management process and from this assign work methods to various segments of the overall project. This is essentially defining work unit delivery options. The project profile is evaluated to help understand how best to approach the execution process and the associated management requirements needed. Several driving criteria can influence the execution decision for the various components. Key among these is:

- How accurate can the requirements be defined?
- What type of status tracking is required?
- What level of risk management is necessary?
- What are the constraints that have to be met? (Time, cost, functionality, etc.)

Answers to these questions will guide the work packaging process.

Previous sections of the text have described the various delivery options as:

- Waterfall—tasks reasonably defined
- Iterative agile—scope loosely defined; prototype acceptable

- ■ Critical Chain—time compression needed
- ■ Modified Scrum—scope loosely defined but physical product

After the requirements are expanded to the appropriate level a WBS will show the required work package execution structure. This view becomes the target for deciding how best to produce those work elements.

Execution

This layer executes the work plan according to the packaging decisions outlined above. There could be four potential work options from the decisions above.

Monitor and Control

The control decision is based on the stated requirements for the project and the formats for this will be selected from a standard metrics list. Even though the list is variable, the formats for this will be familiar. At the lowest level of control, a Gantt chart would be used to show the component level roadmap type schedule. More detailed control parameters would be added as necessary based on the project profile and internal team-level controls will be dictated by the execution method used.

Resource Management

The role of this process is to supply appropriate resources to the projects as agreed upon. Resource staffing plans would be the common communication link between the project team and the resource supplier. Ownership of resources is a common problem in a functionally organized host. Projects often require resources from multiple organizational units which make this a more complex management issue than assuming one resource type owned by one organizational supplier. Regardless of the organizational structure relationship, the resource status between the source and project must be managed. Many projects fail because of this very gap in management. If a resource is not available per the staffing plan, even a good project plan will be of little value.

Process Support Architecture

The issue of host organizational support to the project is much more complex than most project environments seem to understand. Chapter 11 described the breadth of potential support. The question now is how to match those elements to the target project. Communication across the various planning layers is needed to accomplish this. By the time the project has evolved to the Delivery Methods stage, planning is needed concerning what items are required. In a mature organization, this should just require a checklist of services needed. The design goal for the supplying organization is to have a mature support architecture. Failing that, there is a gap planning exercise needed in the Delivery function to assess what has to be applied by the project.

As an example of the host linkage setup issue, the project team is to be paid by the host organization as a service to the project. To do that, charge codes need to be established in the system. Envision a WBS code for identifying task structure. This code is typically used to report various financial transactions in the host accounting system. If an employee charges time to that WBS code, there has to be a linkage established between pay and other financial systems. Similarly, HR, procurement, IT, and facilities may all have to have linkages established so that the project can access them.

An Execution Overview Example

An execution-level WBS model example is shown in Figure 14.6. This illustrates a waterfall/Scrum mixed-mode hybrid design, which is judged to be the hardest work design option to describe.

In this view the majority of the plan is based on a traditional predictive task model; however, there is one string of loosely defined scope tasks labeled 1.2.2 through 1.2.5 that are to be executed according to a Modified Scrum sprint option. The chosen method to execute these work units is based on a MoSCoW scope

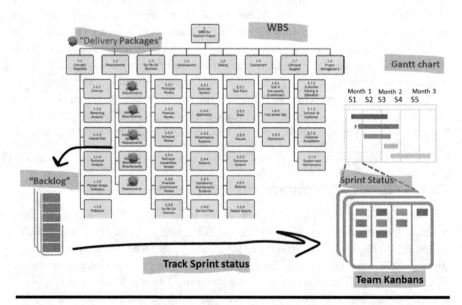

Figure 14.6   Integrated hybrid model overview

definition. All the core tools shown have been described previously. The following is a summary of the execution process modeled by this design:

1. The overall scope is sufficiently defined to produce a task-level WBS based on predictive model assumptions.
2. One string of tasks is segregated in the delivery planning stage to be executed according to Scrum guidelines.
3. Scrum work units are moved to the sprint backlog for scheduling into sprints.
4. Active sprints are tracked using individual Kanban charts to show status.
5. The overall project status can be mapped to a Gantt view with the sprints being shown as fixed bars.
6. Traditional monitoring and control tools can be used for predictive tasks.

This diagram illustrates how the various task work streams are managed. What is not shown in this diagram is the project plan that would be created for the predictive tasks, but that is similar to the typical waterfall plan previously described in Chapter 8. This view would be characterized by traditional control with a sidebar of Scrum sprints to deal with those selected work units.

Work Chain Execution

The concept of a work chain as used here simply means that a group of tasks that are selected for either time compression as outlined in the previous example or to expedite some subset of the overall plan. A work chain can be preplanned during the delivery planning phase or used as a rescheduling tool. This execution option is available for an entire project or a subset as outlined here. Selected tasks can be in the vertical chain of a WBS, selected tasks as part of a sequential chain, or just a cluster of tasks in a time frame. The role of the work chain is to expedite the execution of that task collection. A work chain can exist either in a waterfall structure or even in a sprint. It doesn't matter. This option should be looked at as a management focus item. The definition of a work chain is:

> *A collection of defined tasks or desired features that will be rigorously pursued from a schedule and resource point of view. Within this grouping of defined goals, the team is focused only on those targets and the goal is to accomplish the desired output efficiently.*

In the case of a predictive (waterfall) project plan, a task chain could be shown schematically in Figure 14.7.

Assume that the boxes in Figure 14.7 represent defined traditional tasks with associated resource requirements. In this example, the task list represents more than a single chain., think of it more as a focal point or critical "chunk" of work. This task group can be viewed as a Critical Chain, or simply a focus group of tasks being

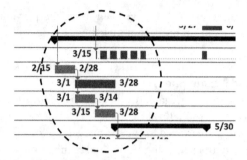

Figure 14.7 Sample work chain

executed under tight management control. If time compression is the goal, the principles of Critical Chain theory are applied. Another way of looking at a work chain is that it offers many of the expedited views associated with the Goldratt theory but the resources required may not be available to execute that approach. Nevertheless, there is value in the formal management focus. It might even be reasonable to suggest that the entire project be broken into groups of this type from a management delegation standpoint.

Modification #2—Maximum Time Compression

The goal is this version is to carry the work chain example to a formal level using the Critical Chain operational rules. The desired goal is to complete the project in minimum time using Critical Chain logic. If sprints remain in the delivery plan, they can be left as described. In this case, resource management discipline now becomes the major focus. The traditional project task plan portion will need to be converted to 50/50-time estimates and buffers added as described in Chapter 10. An abbreviated sample plan is shown below to illustrate this transformation. The steps to convert the traditional plan to this format are:

1. Ensure that all duration times are set to 50/50 (i.e., 50% chance of meeting that value).
2. Insert buffers according to the CC theory.
3. Verify that resources are positioned to meet the defined plan.

If there is a single critical chain in the project, it may be reasonable to just show a single completion buffer. Also, if there should be multiple chain groups involved, buffers will be needed to protect the critical path from the sub-chains overrunning.

A reader is warned at this point. The use of work chains using buffers as described here plays havoc with traditional plans versus actual control schemes. In the traditional plan, the design is to view execution as a bus schedule. Think of it this way—In the traditional plan, you stand at a certain task point and wait for the previous task to complete on say June 4. Remember the route will have extra time, so the bus may need to stop and wait for the schedule—i.e., procrastinate. In this expedited chain view, we now don't know when the bus is scheduled, so we need a warning that it is coming. All we know is that it is coming as fast as possible (down the task chain). This alteration in project control logic will challenge the traditional project manager as well as other organizational levels. Assume this version is designed to be full-time compressed according to Critical Chain principles.

Figure 14.8 shows a simple Critical Chain plan with one critical path and one non-critical path. Three buffers are shown. Note the final task (#11) is a project buffer used to protect an overrun to the plan. This task has a target finish date, but none of the other tasks will have a defined date. Remember, this is a relay race to completion with all of the operational rules defined in the Critical Chain chapter. There is one mandatory buffer according to the theory and that is the non-critical chain ending at task #7. This is called a feeding buffer. To illustrate just one more item in the CC model, assume that task #9 required a scarce resource. To protect that task from overrunning a buffer is placed in front of it. Collectively, this illustrates why the buffer logic could be a hard sell in organizations. It looks like padding is going on everywhere and it is, but with appropriate logic in this case.

The art of buffer design can become complex, and expansion of this mechanic should be done carefully. The task structure is modified by inserting buffers into key places in the plan such that the critical path will be protected. In other words, this project is being managed to deliver according to the modified task chains. The project schedule in this sample is being protected through the three buffers: project, feeding, and resource. Each of these is identified in the task Gantt bar view.

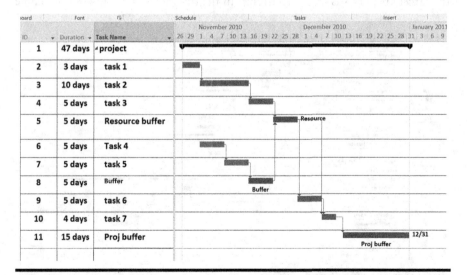

Figure 14.8　Critical chain plan

Most of the scheduling value of the Critical Chain theory comes from the estimating logic and use of defined buffers, along with the related resource management processes. The use of this technique is designed to improve completion cycle times by at least 20%; however, the operational environment will need to be tuned to this process. Resource management becomes the main focus and no multitasking is allowed. Failure to obey the execution rules will negate the value of this process.

Modification #3: Embedded Work Chains

Figure 14.9 shows how a subset of the project plan can be identified as a work chain. This is essentially a Critical Chain but created as a subset of tasks within a broader traditional schedule. Think of a work chain as any task grouping that needs to be expedited. In this example, WBS codes 1.4 through 1.6 are the grouped targets. That segment of the project will be executed using CC principles for time compression.

Modification #4: Recouping a Schedule Overrun

Project management involves more than an orderly execution of planned tasks. For a myriad of reasons, it is not uncommon for some segments of a project to become overdue and badly need schedule recovery. Looking at this problem area as a dynamically created work chain is a reasonable view. In this case, the schedule is badly in the ditch so dealing with the originally planned values is no longer worthwhile. Let's say that the new goal is to expedite this segment of the project. The use of a work chain view provides the focus needed to tackle this requirement. One of the classic ways to decrease a schedule is to add additional resources to the target area. This is called "crashing" in the traditional project world. It can be

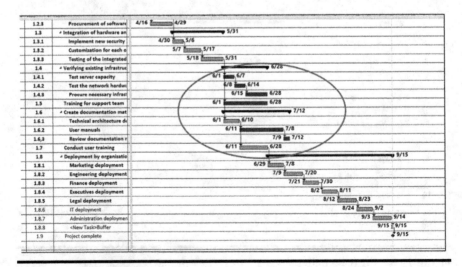

Figure 14.9 Embedded work chains

effective if managed properly, an option to consider is to label the cluster of tasks in the overdue area and focus on improving their status. In a scenario such as this, an experienced official triage manager should be placed in charge of the effort. Every effort will be made to compress the unfinished work and get that segment back on schedule.

Modification #5: Modifying the Scrum Model

Even though Scrum is cataloged as part of the agile model, it has more of a team focus and less of a prototyping aspect. Proponents of this model would agree that it can be used in a non-preplanned scope environment. However, using this approach for a product target necessitates some process modifications to deal with. This work option has been defined in the text as Modified Scrum.

If we start at the top of the integrated model structure, everything stays the same until the delivery planning decision is made. At this point, the WBS boxes begin to take on more of a sprint flavor with MoSCoW requirements, but the boxes still represent work to be produced. The difference that starts to surface at this point is requirements are more loosely defined and graded. A sprint schedule is needed to sequence the work execution. The one attribute that now becomes more obvious is that the ability to produce a fixed calendar schedule is more arbitrary. Each defined sprint would have a certain mushy schedule feeling to it regarding what will occur from the sprint. In the standard definition of Scrum, a sprint occurs in a fixed time box, but multiple sprints are envisioned to complete the requirement In this example, some minimal deliverable is required in one sprint cycle, which means the MoSCoW technique has to be used to define the minimum delivery goal. Here is the proposed Modified Scrum process to utilize this method:

1. Delivery packages are defined via a WBS view.
2. Delivery packages are sequenced into a plana-like network.
3. Various requirements are packaged into a sprint for execution with a scheduled goal.
4. Every effort will be made to deliver the full slate of requirements in the sprint timeframe. When that is not possible, requirements will be omitted in priority order in an attempt to at least obtain the mandatory views. The sprint can be terminated only when that minimum deliverable is achieved.
5. If the minimum requirement is not achieved by the time the sprint is scheduled to terminate, it will be extended with an increased management focus. Additional resources may be allocated if that would be of value.
6. Execution status will be measured using Kanban for internal sprint status and Gantt charts will be used to show a high-level overview of the status.

The use of Scrum in a production environment will necessitate changes in the way requirements must be defined and status tracking will be more based on buffer status.

Monitor and Control Theory

The text has described why the traditional concept of a project plan showing planned versus actual dates for each task is judged to be an archaic and dysfunctional method of status tracking. Recall that the new approach to task estimating destroys this view. Also, scope changes alone in the traditional predictive environment invalidate this approach. Different work types also require different approaches for measuring status. A modified approach to project status needs to reflect this new success-oriented view. If calendar dates are required by management, these need to be presented in probability-type views rather than discrete values that are in error by definition. Figure 14.10 illustrates a more accurate completion forecast than the traditional single value that is wrong. There are commercial techniques that can be used to produce this type of output, especially in the predictive model environment, but similar techniques can be used for hybrid plans.

Beyond the issue of status quantification, it is also important to recognize that project status is now being recognized in a much different manner. Planned product functionality, schedule, and budget are only one view of a project and maybe not even the most important one. Customer satisfaction is emerging as a recognized item. The text described the research project that showed measuring adherence to project goals did not correlate with customer satisfaction three years later. Changing views regarding how to measure project success will significantly impact this area of the model, but we must just leave that as a warning for now.

Historically, the use of buffers in a project plan was viewed by senior management as puffing up the plan. As described here, this is a vital part of the operational

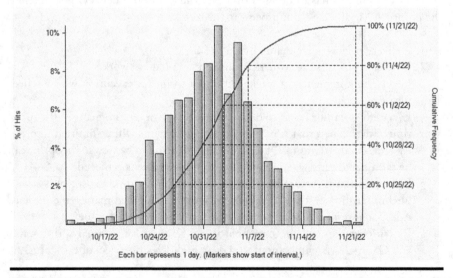

Figure 14.10   Schedule completion forecast. **Output produced by Full Monte from Barbecana

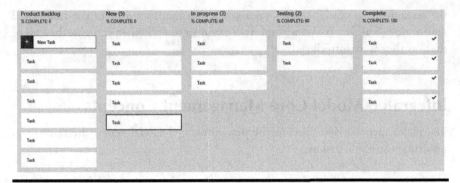

Figure 14.11   Kanban chart

model. Senior management is going to have to be educated as to why this is needed and what it is uncovering in the way of bad practices.

One of the emerging ways of tracking work status at a lower level is through the use of Kanban charts (see Figure 14.11). This is a visual chart showing how work is moving through a process. It can be a sprint or monthly task view in a traditional project.

Choosing the Delivery Option

Each of the classic delivery models described has been labeled as its strengths in delivery. This topic has been stressed from the beginning as a way to match unique project delivery needs to work strategy. Time must be spent in the planning stage to evaluate these differences and the desired delivery goals. What is defined at this stage will impact the rest of the management cycle. Two of the key delivery option decision drivers are the level of scope definition and the specific delivery goals (speed, risk, budget, etc.). The greater the focus on control and assessment, the more likely the favored model will be waterfall based. In the case of software development, classic agile or Scrum has already won that battle even though there are limited domain issues with how to better merge this into the portfolio decision-making layer. Also, the requirement for broader documentation is not well covered in the traditional agile perspective.

One may get to this point in the text and feel that all of this has just been another band-aid to preserve the waterfall model. Admittedly there has been an attempt to preserve as much of the workable parts of all the selected models used to minimize the learning curve, but a review of the items covered suggests that many changes in the management process are included in the new model. The list below contains a summary of non-waterfall items covered in various parts of the text that had non-traditional management concepts applied. The following list is grouped into two parts. The first six items outline process changes for the new model, while the second group of eight items relates to slight modifications in the traditional model view. This list is also a good final exam to test whether you absorbed all of these as they were introduced.

Integrated Model Core Management Concepts

The major changes described in the new integrated model versus the traditional models are summarized as:

1. The new model combines several classic model pieces into the overall plan structure customized from the project profile.
2. The new model provides a more realistic understanding of the overall project management role and the decision structure for an appropriate supporting model.
3. The new model focuses on success process drivers based on industry surveys.
4. The new model uses explicit project profiles to design the management structure around a defined decision block architecture.
5. The new model requires that the project success definition be outlined as part of the plan and this will be used in internal decision-making.
6. The new model outlined gaps found in the classic models and provided techniques to mitigate their impact.

Integrated Management Model Enhancements

The new model expanded the traditional management domain focus in the following ways:

1. The layered decision approach is more logical that the traditional life cycle time view.
2. Recognition of a broader decision environment with EPMO, organization support, and resource components of the model umbrella.
3. Linkage of EPMO to active projects.
4. Use of graded MoSCoW-type requirements for use as a new work delivery strategy.
5. Modified definition for proper task estimating to improve throughput (50/50).
6. Concept of four multiple work execution options within one structure.
7. Provided a flexible management approach for risk and scope definition to minimize excessive planning.
8. Recognition that fixed date status tracking is not an appropriate method.
9. Described a proprietary technique to handle predictive iteration type work (Modified Scrum). In parallel with traditional predictive work.

Summary

This chapter has provided a broad overview of the integrated model architecture and key processes. Decision logic for the various layers was summarized. The processes and block execution decision structure, along with best practices represent key focus guidelines to keep the project moving in the right direction. The management mechanics for navigating this model are based on a classical elaboration theory that best describes how the process should unfold. Each of the decision steps used the stated delivery requirements and matches that with the project characteristics to define how best to execute the work units. The major difference between the integrated model and traditional ones lies in its evaluation of the project profile and associating that with appropriate multiple work delivery options.

Most important, the project decisions are focused on the unique characteristics of the target along with its external constraints.

Before committing to using any of the current models, it is important to define the project profile. Results from this provide management guidance through the life cycle.

The next level of detail provided here does not equate to a technical Users Guide because that implies that there is a single step for each item. Formalizing such artifacts violates the flexibility concept. Layers of the model indicate what kinds of decisions need to be made, but not what a particular decision should be. The goal of the decision process is to eliminate non-productive work and only do what fits the target project goal structure.

Reference

PMI. 2017. *A Guide to the Project Management Body of Knowledge (PMBOK Guide)*. 6th ed. Newtown Square, PA: Project Management Institute.

Chapter 15

Modified Management Processes

Previous sections of the text have described both success and failure-oriented management processes. Chapters 13 and 14 described the underlying component management structure logic for an integrated decision model. This structure defined a decision layer view of life cycle management decisions that utilize a project profile to customize an appropriate set of decisions that align the management approach to produce successful outcomes. This chapter follows this with a description of selected processes that are different from traditional approaches. Even though the integrated decision structure looks similar to traditional life cycle views, multiple processes within the new structure need to be understood as to their reason to exist and their mechanics. Also, these must be utilized as described to obtain the deliverable value advertised. Some will resist these changes until they understand why they are being introduced. The use of general management best practices has been stressed in the text and is summarized here for the more modified processes. Think of best practices as the "doctor, doctor my arm hurts" example from the text, meaning that if an existing process is not working, don't do that. As a group, these are important elements of successful delivery. Many problems in current project practice can be traced to a lack of understanding of the impact these have on negative outcomes. A summary listing of these is included in Chapter 18 under the title "Success Recipes."

Conversion of project management practice to the new model will not be casual or transparent. The various new processes were summarized in the previous chapter and those will require a more analytic approach to the decisions. For that reason, each of the new processes must be understood by the team and all stakeholders. The new integrated model is not contained in a large notebook of steps. It is more

DOI: 10.1201/9781003431091-17

focused on concepts that mitigate failure sources and improves the likelihood of delivery success. The silver bullet in this case comes from the knowledge and understanding related to a management process that better aligns those processes to the target initiative's characteristics and goals. This approach maps the proper work process to a defined project type and that requires more flexibility in the decision steps. The sections below will outline why these new steps are important, or why the traditional management approach needs to change.

Project Profiles

Chapter 2 described the concept of profiling the project to recognize its uniqueness. The model design thesis is that all projects have characteristics that should be used to define the related management process. When viewed at a low level the profile parameters include not only the characteristics of the project but also the environment in which it will be developed. Environmental factors might include limited technical resources or unproven technology. Each of these classes of parameters can affect the delivery strategy in different ways. The point with each is they need to be taken into account in the decision process. In addition to optional decisions related to multiple work execution, the prioritization of delivery goals also impacts how various processes in the model should be defined.

Risk Strategy

From an engineering model perspective, there are documented risk assessment models to guide one through a rigorous analysis of potential unknowns that can affect the project outcome. Avoiding potential risk events is a positive goal, but this process can be time-consuming and still not effective. If the project is costly, has a highly competitive potential, high technical risk, or other potentially negative factors, there is an increased need to formally evaluate these sources. The dark side of this decision is that it is time-consuming and expensive and may well fail to identify the unknown event that will later produce a negative outcome for the project. A review of real-world risk event examples reveals how such events can sabotage or even ruin the value of a project. Although a highly important aspect of project management, the risk management process is the least mature of all defined processes. However, that does not mean the questionable accuracy of a formal assessment should be avoided. Some reasonable risk assessment is required in every project, with the real decision question being how detailed should this be. There are many factors associated with this decision and a single prescriptive answer is not reasonable. The one prescription that does seem obvious is to openly discuss this aspect of the project and attempt to deal with these unknown items before they surface or become worse.

Two required organizational techniques should always be employed related to this topic. First, develop a risk culture in the organization by sensitizing all to the impact of this fuzzy class of negative events. As an example, the risk of fire in a building can be often managed by education and various mitigation techniques. The same approach can be taken with various high-potential risk items, with or without formal evaluation. A recommended companion decision to this area is the assignment of a formal *Risk Owner* to various segments of the project. Their job is to be the leader in reviewing the environment for items that could impact their area. The combination of these two sources should be considered universal recommendations. Murphy's Law says "Anything that can go wrong will do so at the worst possible time." This is the attitude that the project team has to have regarding project risk. Humans often tend to ignore the concept that something can go wrong if it doesn't occur frequently. Also, be sensitive to anticipating that some project risk events will come from unusual sources. Having a risk culture is an important component of successful delivery.

Scope

There is one unchallenged truism about project scope. That is, the initial delivery specifications will significantly change during the life cycle. In many cases, the users do not know what they want or will change their minds after they better understand the target goal. Based on this assessment, it is not hard to conclude that scope management is the Achilles heel of project management. It is very hard to build a coherent plan to produce a defined outcome for something that will change in some unknown way later. It is this characteristic that has a significant influence on defining the best method to attack the management process.

One of the major decision variables is the degree to which the project scope can be accurately defined. The belief of a clear definition more likely leads to a predictive style of work execution strategies as outlined in the waterfall model, while lesser confidence tends to move the execution strategy into either heavy change requests or possibly to the iteration style of execution. Related to this decision the type of work streams that fit the scope stability. Also, it is important to recognize that most projects have mixed scope areas, so there is a high potential that multiple work streams should be used if one is sensitive to matching this characteristic to the required deliverable.

The traditional approach to scope definition is single-valued, which leads to functional specifications that might describe the design deliverable goal to be 100 mph and weigh 50 pounds. Conversely, the integrated model has highlighted the value of grading specification goals into categories, with maybe only the highest-level ones being required for successful delivery. This has the potential advantage of defining work completion differently and improving goal outcomes. Trying to deliver a complete list of delivery specifications often includes items that are not necessarily needed, but desirable to have. The graded approach also fits the Modified Scrum work option, which may well turn out to be the most popular future work option once the process is matured. This modified approach to redefining goal deliverables also has the potential to increase the usage of more flexible scope values, which allows improved flexibility in choosing work strategies than offered in a traditional predictive scope environment.

Task Estimating

Traditional approaches to task estimating are based on padding the raw estimate to try to make the planned value match the actual outcome. Chapter 10 described the psychological effects of this strategy and showed how it did not work as logic would suggest. More importantly, it brings many bad practices along with it. To create an action-oriented work environment this practice has to be scrapped and redesigned using what the text calls the 50/50 model. Like many things, this change causes other processes to require redesign as well. This is the highest redesign modification recommended and the one that has the biggest potential for improved delivery times. All projects have some inherent need to complete in a reasonable time, and this point has been somewhat lost in the traditional waterfall model view.

There are many techniques documented in the literature to produce project and task estimates. One should become knowledgeable in these as they are designed to produce what one might describe as the "raw estimate," meaning this is what would be expected given the work specification. The thing that happens after this is the source of the problem. Knowing that an estimate is not 100% accurate but still likely to have variability, it seems logical to just add a little to protect the estimate from overrunning. Measured overruns are viewed negatively in the traditional project culture. Review the Goldratt theory outlined in Chapter 10 to see how this approach does not accomplish the desired outcome or culture. Using this logic, tasks will still overrun and the project schedule has lost all integrity.

50/50 Process

Factors that impact the ability to produce exact task estimates include clear work definition, the skill of the assigned worker, and other environmental items such as weather that can play into the variability. It is culturally difficult to understand the 50/50 rule when even current padded estimates are being overrun, so this topic will require quite a bit of conversation. For groups that have been criticized for overrunning, this will be a doubly hard argument. The Critical Chain model theory provides a strong case to modify this process as described. Converting to this modified approach is one of the strongest recommendations of the text. Remember the behavioral factors that dominate this issue:

- Student syndrome—delay starting
- Procrastination—fragmented view of work to be accomplished
- Multitasking—loss of productivity
- Parkinson's Law—work fills the time available

Once one understands this set of factors and the logic for modifying task estimates, it becomes much more motivational to work on changing the "variance culture." If one analyzes this logically, the goal of a project is not to make numbers come out right, it is to produce planned items in an optimum time. Padding tasks violate that goal and create other poor internal behavior that negatively impacts the outcome.

This 50/50 approach energizes the work task process to minimize wasted time, but it admittedly also creates more task overruns as a byproduct. In practice, it may be more acceptable to use a slight modification to this ratio, but something along this order is required to produce the desired work culture. Figure 15.1 shows the probability view between the traditional view and the modified approach.

The traditional view would use the 100% estimate, while the modified view would use the more restricted option—most likely (50%) or a 60% value. Either one of the modified options moves the task estimating process in the right direction to avoid procrastination and schedule delays.

This processing strategy creates the environment that is desired and that is improved action-oriented task completion. In other words, this is the method to improve task cycle times even though it brings some other complicating factors along with it. Successfully implementing this practice has a major impact on project completion, but strikes at the heart of other marginal practices in the current environment such as a culture of procrastination and multitasking. In essence, both of these practices fall under the umbrella of poor process discipline. Tasks that need to be worked on now must now be focused on with high priority. The metaphor

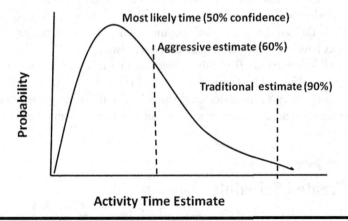

Figure 15.1   Modified estimating view

used in the text is to view the execution sequence as a track race. Even though this is viewed as a positive, we need to look at some of the negative aspects of this decision.

Resistance to the 50/50 Rule

The project is now running at warp speed and the only remaining concern is to finish at some buffered completion point. The 50/50 estimating strategy will almost assuredly produce task overruns, and some type of buffering strategy is needed to protect the project schedule. However, buffers are also looked at negatively. Historically, buffers known in project plans were viewed negatively since they looked like padding, which they are except in this case they are based on valid logic and kept isolated. The traditional process simply buffered every task, which bloated the overall plan while essentially hiding where the real overruns occur. Separating the buffering from the task will reveal what is going on with individual tasks. Showing a large completion buffer to protect the project schedule will assuredly open up discussion on the legitimacy of the buffer value chosen. This will necessitate an educational effort for various groups to explain why this method is necessary and why it is a solid technique.

Plans created with modified task estimates and buffers change the geometry of status tracking. One way to view the 50/50 tasks' status is that they are now floating in space with no defined start or complete time. For a project with one critical path and one competition buffer, there is no fixed end point except the one shown at the end of the project buffer. The relay track race is a great metaphor for thinking about this. Envision the active task as a track runner going as fast as possible. When they get to the end of their task, they hand it off to the next runner, and so on through the chain. Note that you cannot accurately evaluate how the race is going (unless you have run a race like this many times). The only active status variable for this race is the buffers inserted into the chain. In this case, it is the completion buffer. One status measure used might say that if you have accomplished 70% of the estimated chain time and consumed only 50% of the buffer, things seem to be within bounds. Even though that is a typical measure used. The flaw in this assumption is that all tasks have the same performance characteristic.

As a final summary on this point, Task padding is one of the worst practices in the traditional model and must be curtailed if one wishes to improve delivery outcomes. All parties must understand the role of buffers in this new view. This will be a management challenge to accomplish, but the performance improvement potential is clear.

WBS Created Schedule

A WBS has multiple roles in the new model. important one is translating the task estimating values into an initial project schedule. Figure 15.2 shows a sample

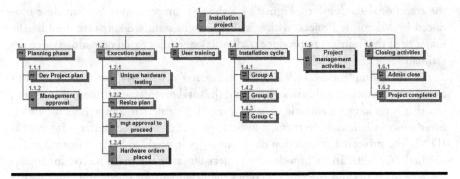

Figure 15.2   Estimating example

project WBS. Assume that all of the lower-level boxes represent work elements that have been estimated using the 50/50 rule.

In this view, the boxes can represent either predictive tasks or iterative sprints. The box estimate represents a sprint time box or predictive task that needs to be delivered. Chapters 8, 14 and igure 14.2 previously illustrated how data from this type of scope overview evolves into an initial project plan structure, so we won't repeat that mechanic here. That example also showed a project buffer used to protect the completion date. This same process also produced a Gantt-type bar schedule for each defined box. All of this process works fine until we recognize that these are 50/50 estimates and are likely to overrun. This raises the question of what is the bar trying to show if it is not a schedule. In the future, the Gantt view will at best be a roadmap of task sequence but poor in terms of defining schedules for tasks. Here we have resolved one of the most troubling project management issues and created another visible inaccuracy that will require changes in the way project status is viewed. The concept of a planned start and finish is now recognized as invalid. We now need to deal with that issue.

Status Tracking

If the management goal is to track the project's status using the new integrated model, the traditional approach will no longer work as the traditional model perceived it did. As described previously, the traditional view of defined task start and finish dates is an illusion. It did provide target dates, but they were not appropriate in that they are excessively padded. The new model approach has attempted to better match the estimate to what might be achieved. While the new estimating process is valid from the factors outlined earlier, it does corrupt the classic Gantt bar schedule view, even if the bars are computed using valid network scheduling mechanics. At first view, this point may be hard to understand, but the key is to view task duration more as a probability distribution, rather than a single value as the traditional model does. Figure 14.3 shows a simple Gantt bar schedule produced by Microsoft Project, which is the de facto standard software used by the majority of the industry to produce traditional schedules based on padded duration estimates (see Figure 15.3).

In this example, each task is 50/50 estimated at ten days duration. The traditional interpretation of this would say that the total project cycle time is thirty days, but that is no longer a realistic view with the 50/50 rule. To compensate for anticipated probabilistic task overruns, a five-day project completion buffer is inserted at ID #5. The projected completion date is now "estimated" to be 1/10 instead of the original 1/3 value. In this mode, the project buffer is the main metric for signaling true overruns and therefore becomes the new focus for status and control. The only dynamic quantitative tracking variable is now the buffer status as tasks are completed and begin to consume (eat) the buffer. Furthermore, the validity of the appropriate buffer size will be hard to verify until the project is completed.

A typical question related to this process involves how to define the proper size for a buffer. There are multiple ways to do this and none are universally accurate. A statistical quantitative method for this would be to simulate the project tasks with variable sizing parameters such as found in the three-estimate classic PERT model (early, most likely, and pessimistic). This technique along with computer utility software such as Barbecana's Full Monte can provide insight into potential range values for the project (Barbecana). This simulated schedule distribution profile view could then be used to estimate an appropriate buffer size. Other less sophisticated empirical methods are probably adequate in most cases. Experience with a particular project type would the estimating accuracy over time. Also, Critical Chain related authors have published research papers on buffer sizing and these may be of help as well as calculating buffer sizes. Regardless of the method used to size buffers, the process of evaluating project status is completely changed from the traditional fixed calendar approach.

As tasks overrun (or underrun), the corresponding buffer is adjusted to show how much variability is left in the target date. One status interpretation is to compare the percent completion of the project tasks versus the amount of buffer consumed. For example, if the project is 50% completed, the buffer should be no more

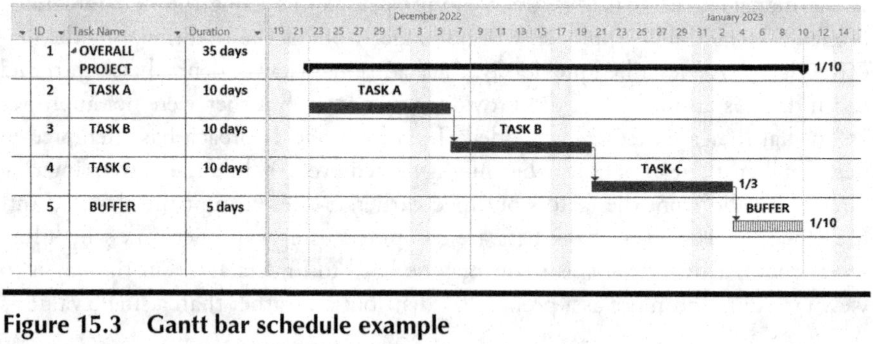

Figure 15.3   Gantt bar schedule example

than 50% consumed. This view is called the buffer "burn rate." The example shown here uses only one project buffer. If the full Goldratt model theory is followed, the approach to buffering is much more complex than described here, but the status process approach outlined here is similar regarding how the buffer burn consumption is interpreted. There will just be more buffers to interpret. The recommended goal here is to keep this model concept as simple as possible so the recommendation is to start with only a project buffer, and then decide if a more complex view is justified.

Resource Management

There is yet one other added management process focus that is highlighted by the new estimating process. A traditional project plan implicitly views resources as available and waiting according to the bus schedule; however, the revised estimating process adds a new dimension to resource management. Since there is no fixed time indicated for when a resource is needed, the approach to managing resources must be more dynamic. The CC model outlines what this requirement looks like, but in essence, it says that there needs to be a visible resource alert system to move the next resource group into place before prior task completion. This is an increase in resource discipline above the current view. Resources are often not in place per the schedule, and this is a common source for schedule overruns. Regardless of the management model used, gaps in resource availability represent a clear source of schedule overrun. The new model requires an increased focus on resource availability, so there must be some formal management process to focus on this as it represents a major success factor in its own right. In examining traditional projects, one gets the feeling that this is not well understood. Regardless of the management model used, the timely availability of resources is a universal goal. Not having timely resources in place sabotages even a perfect schedule and this is often the result. Principles outlined in the Critical Chain theory are good process guides for this role.

Traditional Project Tracking Myths

This chapter has already done a lot to critique some of the common traditional performance gaps processes and status tracking is just one more example. Various previous model descriptions have pointed out that the use of padded fixed task schedule dates does not fit reality and this practice has the effect of slowing the project down and does not match reality. The task scheduling result has been labeled as the bus schedule. The key point to understand here is that traditional planned versus actual tracking does not do what it appears to do and that is to show when a task is truly overrunning. Here is a list of false beliefs regarding this topic:

■ Traditional task's planned start and finish dates do not reflect realistic dates for the task, but rather a bloated date that still has a high potential to overrun (see Chapter 10).
■ The defined padded dates do not motivate speedier task completion, but rather create a stagnate work culture.
■ Resource management processes are not typically tightly triggered to the task execution process, which further creates schedule gaps.
■ Plan versus actual comparisons does not represent true status.
■ The traditional plan completion date is bloated and will still likely overrun, so its structure is not considered a valid status indicator.

Two critical management issues emerge from this new approach to task estimating and status tracking. First, the communication of completion dates must now move to a new approach that better reflects reality. Task overruns are now the norm and will need to be somehow protected by schedule buffers, which changes the proper interpretation of what task completion is. The traditional view was inaccurate because of the bloated values used, and the new approach must be viewed more like indeterminate size tasks floating in time, so fixed calendar dates are no longer relevant measures. The primary enticement of the new approach is that it opens task execution up into a more actionable approach that offers real potential for schedule compression. Recall that the integrated model made special mention of a resource component. This is a clear reason justifying why such recognition is a valid management activity. In the task execution scenario, the key is to alert the appropriate resources that a task will be completed. One mechanic for this is to signal the completion date three to five days in advance. The resources should then work to prepare to start work on the next task promptly. This is another not-so-obvious way in which projects lag in execution.

Plan versus Actual Measures

The traditional project plan contains both the start and finish time for every task, so control measures focus on variations in these values. The schedule illusion here relates to the task padding that in essence destroys the integrity of the estimate. This type of comparison is not measuring variances at all. It is measuring the worst-case estimate for the task with its bloated values. This evaluation approach is not only invalid in its interpretation, but it stimulates a "go slow" culture because of the extra time loaded into the task estimate. In the 50/50 approach, team members are now encouraged to go as fast as they can as there is no slack built into the estimates. Management will have to understand that this planning approach is attempting to compress the time frame and the visible buffers are necessary to handle the overruns.

The myth of showing invalid dates that allow poor performance to occur is now gone and that is a good thing. The fact that the project is moving faster now but we can only evaluate how fast once the task is finished may be unsettling to the high-level reviewer, but that is a better approach. The track race metaphor is real here. If task chains are being executed as described, the buffer burn rate will be the best indicator of status.

Tracking Hybrid Work Queues

The existence of hybrid work queues further complicates the approach to communicating project status. Chapter 14 described how multiple work queues could be created for a dual predictive traditional task and a Modified Scrum set of execution sprints occurring in parallel. The existence of parallel work queues creates differences in not only the work mechanics but also techniques to track the status of these. As combinations of work options for tasks as chosen, such as Critical Chain or Modified Scrum, the physical status view of these can be represented by a bar on a Gantt chart without associated calendar dates as described above. The iterative work units will be shown as manually scheduled fixed-time boxes.

If the entire project uses the agile or Scrum model with iterations, there is a question of how to "schedule" termination since the true definition is when the customer is satisfied. That has a nice sounding ring to it, but in the real world probably not what one would call tight control. Most likely, there will be a management constraint placed on the number of sprints or budget and this may well be the defined task completion. Within this constraint boundary, the iterative process would execute according to agile principles. Figure 15.4 shows a combined agile and predictive execution plan. Two tasks have been identified to be executed via the Modified Scrum sprint method using graded (MoSCoW) type requirements. Both of these are extracted from the predictive task list and moved to a sprint workflow queue where they will be executed using appropriate sprint principles.

In this schematic example, the agile component has two defined sprints with a constraint defining the required stop point. The assumption made here is that

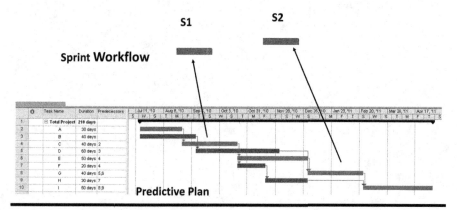

Figure 15.4   Combined agile and predictive work queues

the defined iterations contain estimates suitable to produce sufficient delivery "features" for the planned output. Time permitting, the sprints may produce more but will be obligated to reach the minimum before termination. Note that an overrun may occur in this situation.

Summary

The modified processes described in this chapter need to be understood to realize the improved value of the integrated model. Understanding that the flow of task work is not a fixed calendar schedule but more a probabilistic box. The track race metaphor will have to be understood in the management process rather than the traditional illusion that defines an artificially fixed date that does not equate to reality. Techniques for status tracking and productivity will need to be modified along the lines outlined here. The use of buffers in the project plan will create questions about their legitimacy. This will be just one of the implementation challenges converting from the traditional simplistic model that has long delivered false views of status.

Reference

Barbecana. 2022. Full Monte. https://www.barbecana.com/full-monte/ (Accessed December 8, 2022).

Chapter 16

Integrated Model Tutorial

The goal of this chapter is to outline the process related to what work execution model to use and how to manage multiple work queues. More specifics regarding the criteria for selecting a particular execution approach and some of the attendant issues that come from that selection will be covered. Also, some basic processes are recommended for use in all of the models.

The integrated model structure and related processes offer management decision guidelines that aid in successful delivery. This includes both the decision process and various success-oriented work strategies. The various text chapters warned that the resulting integrated model approach was not a deterministic cookbook and that remains the case. At this point, the background theoretical logic of the overall model has been shown, along with selected work execution optional practices that are associated with it.

Recommended Success Drivers

Each of the processes outlined here has been rationalized regarding their success role. Improper management of each of these will negatively impact the future project outcome.

Requirements Definition. There is improved execution flexibility if requirements are defined in a graded MoSCoW format, rather than single discrete values. The accuracy of the work requirements definition is described as the Achilles heel of the predictive model and one of the major positive attributes of the iterative approach.

DOI: 10.1201/9781003431091-18

Project Profile. The management model should be mapped from the characteristics, delivery goals, and constraints of the target project. This profile should be explicitly defined before outlining an associated management approach.

Scope Management. Excessive changes in deliverable goals can make the management process more complex. Use a scope reserve to fund approved changes and aid in tracking the level of change. When deliverable requirements cannot be defined adequately, the use of iterative techniques can be helpful. Above all, it is vital to have active user involvement in the execution process to provide timely feedback.

Risk Management. The approach to handling project risk is subtle and complex. First, don't ignore it. The recommended strategy is to balance time spent on formal evaluation versus the perceived risk level associated with the project. Formal risk assessment can offer good insights into this problem, but can also require significant time. The process goal is to balance the level of assessment to the problem. Consider using formal risk owners for tactical review and implement a risk culture for the team.

Task estimating. Extensive background on this topic is scattered throughout the text. The basic rule of thumb is to establish a culture of 50/50 estimating and timely resource availability. Keep the Critical Chain theory logic in mind during project execution.

Critical Chain Theory. There are many worthwhile concepts found in this theory; however, the full use of its buffer logic may be too excessive for beginners. Most of the value from this approach is found in task estimating, resource management, and project or stage buffers.

Project Success. The definition of project success can be more than defined by the traditional model. The working definition should be formalized by ranked parameters, and this result then can be used in making tradeoff decisions.

Management Tools

WBS

The Work Breakdown Structure is the fundamental packaging tool to describe how the project is elaborated from approval to completion. This can be used to identify phases, subsystems, work groups, and tasks. This view intends to describe how the project is to be managed. At the lower level, various work units are labeled for execution by one of the four optional types. A project delivery plan can be produced from this view as well. The WBS is considered a core communication and planning artifact.

Gantt Charts

Gantt charts are so engrained in the culture of project status that they would be hard to omit. Regardless of the technical flaws related to this tool, it is kept in the model to show at least relative time frames or status for execution and work sequence. Beware that some of the time compression strategies have the effect of making work unit schedules more of a floating variable based on performance so this impacts the typical use of Gantt schedule views. Note the Kanban example later in this chapter that uses a combination approach with the Gantt view. This further illustrates how this tool can be used for communication purposes.

Work Execution Models

Predictive (Waterfall) Model

The ability to define task requirements accurately hampers the effective use of this model. Its primary value is in its simplicity, but typically does a poor job of forecasting future outcomes and does little to motivate accelerated completion times. This is the most mature model in the option set, but not understanding and properly dealing with the base assumptions makes it a marginal choice for successful delivery. The text chapter sufficiently outlined those assumptions and the issues with this model.

Critical Chain

This model requires both the discipline of a waterfall model and a mature management culture that can understand the role of complex buffers. The model is very specific regarding how it works and the behavioral factors involved. For that reason, it is an important element of all models. Full use of its library of buffer logic and related status control makes this a very complex undertaking, but the model is valid in concept. If time compression is the goal, this approach offers the best set of processes. The mechanics regarding task estimating and resources provide valuable insights for all projects. The text chapter provided a good overview, but more detailed background sources can be found on the Internet and Chapter 10 references.

Agile Principles

The basic agile principles were outlined in the text but operationally other dialects such as Scrum have matured these initial views into multiple sub-process groups. The Scrum model will be used to describe this as an operational management approach for predictive-type tasks, while the basic Scrum approach is the most popular model for standard iterative efforts.

Modified Scrum

This is the most used version of agile principles and may well still be evolving. The use of sprints for work execution and Kanban for status tracking are the core elements.

Hybrid Work Structure Example Plan

Figure 16.1 shows a traditional WBS for a sample project. This represents the defined work units aggregated into major work groups.

At this stage, the work required has been identified into either traditional work packages or Modified Scrum work units (although those designations are not shown on the WBS). Previous chapters have described how this view can be converted into a traditional project plan by adding work unit duration and sequence links. We won't repeat that mechanic here but assume that is done. From this first-cut view, decisions are quantified regarding how the project is to be executed. The following are four key components of these decisions:

1. Based on the scope decision, define the execution option that best fits that decision.
2. Specification of the level of scope definition for each work unit using the recommended MoSCoW approach for scope definition to support Modified Scrum logic and sprints.
3. Use the 50/50 estimating technique as the recommended scheduling method based on the Critical Chain logic.
4. Insert buffers into the plan to protect key mid-stage points and the completion date.

Translating these decisions into a work-related view produces the project plan shown in Figure.

This example is designed to illustrate both format familiarity and flexibility in the integrated model. Various bar color coding can be used to help understand the various work unit types such as sprints, buffers, and milestones (see Figure 16.2). Traditional project managers will easily see the structure of the project and what

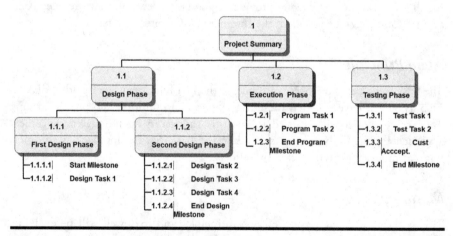

Figure 16.1   Sample project WBS

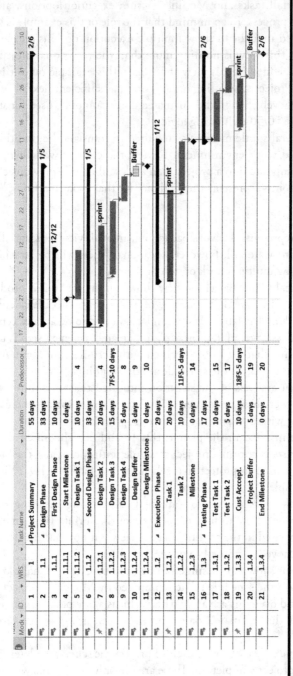

Figure 16.2 Project plan

appears to be a traditional schedule (although it is not). This example shows a combination of waterfall tasks and Modified Scrum execution options, all using the 50/50 estimating process. Keep in mind that the role of this example is to illustrate how the integrated model can show multiple execution options and buffers. One might argue that this view is nothing more than the traditional project plan, but there is more embedded in this view than that. This example output is produced using Microsoft Project scheduling options and has sufficient flexibility to be used for a project plan, even though the calculated dates cannot be used for status tracking because of the 50/50 estimating logic.

There are several subtle mechanics required to translate the traditional project plan into a mixed work environment. Review the defined work items shown in Figure 16.2 that migrate into a project plan as described in a previous chapter. The following set of points represents the guidance for that evolution from WBS to task definition:

1. The WBS represents the work packaging, and this view is essentially the same as a standard waterfall; however, additional labels will be attached to these units to show multiple work option decisions.
2. The graphical Gantt bar chart view shown in Figure 16.2 is produced using Microsoft Project to take advantage of its network schedule calculation logic for a base plan. The label "Task" can be assumed to mean any type of work.
3. Once the classic waterfall project plan is created, the next step is to modify it for work units that are to be executed with Scrum sprint logic. Also, buffers are inserted to protect the schedule at critical points.
4. Assume that analysis of the situation indicated that work units with IDs 7, 13, and 19 were best performed using Scrum sprint logic—these are marked on the plan with different bar styles and sprint labels.
5. Buffers are inserted in the plan at IDs 10 and 20. The milestone shown for ID 11 represents a formal project schedule status point and the buffer shown for ID 10 is used to protect that schedule. The buffer at ID 20 is the classic project completion protection buffer.
6. Sprint work units are set for "manual" scheduling meaning they can be moved through time to fit the character of the sprint. This same scheduling logic can be used in the traditional plan view but now has the flexibility to be manually moved as needed to fit the overall plan.
7. All durations shown are assumed to be sized by Critical Chain logic (50/50), which means the scheduled dates computed by Microsoft Project will not be accurate and overruns are anticipated. This is a key point! There is no expectation that the calculated dates shown are meaningful for status-tracking purposes.
8. The example only uses two buffers to protect the scheduled dates—the design stage and project completion. These are the only two points where a calendar date has meaning.

9. Milestones are defined in the plan at UDs 4, 15, and 21. These can be used to provide a measure of status at these key points, but dates shown for tasks should not be used for status analysis purposes. Review the Critical Chain theory for this background logic.

10. All time-restricted tasks in the plan are now floating through time as their actual durations are not expected to be as shown by the Gantt bars. One high-level status approach would be to track the actual completion dates, and this can be done using the software shown here. Given the philosophy of the Critical Chain, these dates would not be used for status tracking.

11. The traditional planning view for work units is to specify the sequence and assume that task estimates are fixed in size. In this case, the task estimates are variable and the sprint logic may be manually moved around to accomplish a needed goal for that sprint process. So long as the manual task fits into the overall structure, this is doable. The management of Scrum sprints can be time variable, and it may be a better reality match to show some units with a manual flag to better reflect how the sprint is to be executed. The three sprint units are set to be manually scheduled in the example. This allows the task to be moved to any time slot for execution so long as it does not affect the linked follow-on task.

All of the comments above illustrate how the alternative execution approach is defined in the plant. The flexibility of this approach is designed to better match the project profile. Scrum work units represent a parallel workflow to the traditional units.

Hybrid Execution Case Study

This example will review the key success driver process and three classic delivery models, and then look deeper into Modified Scrum mechanics for segments of the overall project. In this example, the project plan is first translated as a predictive view, and then modified to accommodate three defined Modified Scrum internal work units. The related operational mechanics for these sprints are then described showing the dual workflow plan.

A second set of mechanics illustrate how to use the MoSCoW logic to plan and track the status of the Modified Scrum. Once defined, the sprint schedule can be moved into a time position that allows it to perform in concert with the traditional tasks in much the same way that a standard predictive task operates. For this example, the project plan's phase work structure is comprised of three design tasks. One of these is planned for execution via Modified Scrum sprint logic, while the remaining two will be executed using traditional predictive tasks. The buffer, at ID 10, is in place to protect the project phase and its formal target milestone (ID 11) from overrunning. The goal here is to drill down into Design Task 2, ID 7, and examine the Modified Scrum logic that could be applied to that sprint. The logic for using this option is based on the notion that the team is knowledgeable and would normally have previous experience with this type of deliverable. This is a tangible type of product with physical output requirements, and a set of graded MoSCoW requirements for each major deliverable attribute is shown in Figure 16.3.

During the earlier planning cycle, the scope definition for this work unit was loosely defined much like standard agile "features" and essentially represented by the equivalent MoSCoW definition. This is considered to be the iterative definition format to drive a Modified Scrum sprint. From this, the experienced sprint team is charged with determining how best to produce the item. This approach is very similar to classic agile, but it is important to understand the subtle differences outlined here.

To satisfy the minimal technical product attributes, the sprint team is challenged in this case to deliver both the "Must Have" and "Should Have" requirements, but only the "Must Have" level is required to terminate the sprint successfully. Note that different minimum delivery values can be defined for each of the attributes. This is viewed as a technical decision and once again represented the management flexibility of the new model. The defined work package is estimated using the same 50/50 rules outlined for other tasks and the sprint is scheduled for a six-week duration to deliver a working product widget. A Kanban chart will be used to review sprint performance measures, and it will be produced every two weeks along with an assessment of achievability to deliver the required output by the end of the sprint cycle. The internal sprint team will react to these measures for ongoing work strategy.

A Kanban board is used to track the sprint status. There are multiple ways to format this view. It can be much like a mini-project status for the life cycle, but an alternative way is to show the improvement in output technical status through the sprint cycle. Figure 16.4 illustrates a very interesting format view to highlight how the technical performance might be displayed.

This format uses the logic of both Kanban and Gantt to display periodic status and once again represents the flexible usage of our core tools. In this case, the desired delivery target is "S" and the deliverable minimum is "M." The tracking chart clearly shows that the sprint is viewed as at least minimally successful. By

Unit of Measure	Must Have	Should Have	Could Have
CFM (flowrate)	20	17	15
LBS (weight)	50	40	30
KW (power)	1	.8	.5

Figure 16.3   Graded requirements using MoSCoW requirements

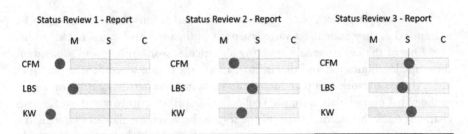

Figure 16.4   Kanban/Gantt performance tracking

the third review cycle, all performance measures are achieved except for weight (LBS) which is very close to the "S" level. This would show deliverable success for the sprint.

The length of a Modified Scrum sprint will often be longer than the typical agile iterative type since a minimal deliverable here has to be produced in one cycle. Output status should be reviewed in at least one- or two-week cycles. In this example, it is decided to formally measure output results every two weeks. The startup steps involve estimating the work required, defining the sprint cycle time, and melding that into the overall project plan as shown in Figure 16.2. Use that view to review how this sprint was linked to the predictive task structure. Also, note that a phase buffer is allocated to protect the formal milestone shown for task ID 11. If this sprint were to not achieve the defined minimum deliverables by the scheduled endpoint, it would technically have to continue with an overrun with the impact on the overall plan similar to a traditional task overrun.

There are two subtle management points embedded in this example. First, the initial planning was much less than traditional. Second, most of the planning and control activities occur during execution and are primarily in the hands of the sprint team. There is a significant delegation of responsibility moved to a lower level. Both of these management concepts are related to the agile success model and directly attack a weakness of the predictive model.

Summary Evaluation

This example provides operational detail insights into the dual workflow logic and how that approach better deals with one of the critical gaps in the predictive model. In this example, there are essentially two work design choices for Scrum units—get as much done as possible in a one-cycle sprint fixed timebox with variable delivery scope, or alternatively work on a classic iteratively defined "feature set" and evaluate the results for further work. The one-cycle sprint must produce at minimum the "must have" portion of the requirements and an overrun is possible, which may require a protection buffer if that work is on the critical path. The design logic of the integrated model is to evaluate how best to execute the specific work units.

In the sample project plan, the impact of these various decisions needs to be recognized as they each affect various aspects of the work and status-tracking process. One of the key drivers in selecting a particular work unit option is based on the scope level question and this in turn affects the future schedule interpretation logic. If the full work unit plan for a project was iterative Scrum, one might choose to show that overall plan as simply a collection of sprint Gantt bars and use Kanban charts to track the internal status of the sprints; however, the flaw in that logic is a belief that even this type of project has some high-level initial planning that would be best shown as fixed tasks with possible buffers to protect the overall schedule. The goal of the integrated model view of work status is to match reality to the related work and not show erroneous calendar dates that have no logical meaning.

There is one more reality-based comment. As indicated in the text, the impact of scope changes and triggered risk events is messy regardless of the model. Recognize that neither of these events is in the initial work plan as they are not defined and may not occur. However, once they are defined during execution, the challenge is how best to show them in the plan. At this point, they represent work that has to be accomplished. In this scenario, Figure 16.2 should now be modified to show this additional work. There is no standardized method to handle these scope extensions to the original plan, but failure to represent them leaves the project adrift in terms of its work requirements for completion. Reserve accounts can be established to cover the cost aspect related to these decisions but the schedule portion can be more administratively difficult to show. Ideally, the plan would be edited to include the new work and then used as a roadmap for ongoing effort. An ideal but administratively complex method of showing these changes would be to use color-coded changes to the Gantt bars and a related data store with details. This area is a problem for the traditional project and will be similar to the new model. This is a reminder that scope change is one of the more difficult aspects of managing the project.

Sprints can be made more visible in the project plan view as coded boxes as shown in the sample project plan, but the initial scope planning view of this needs to be recognized as well to avoid excessive work there. If the designed sprint logic is to have three sprints for the example phase, this would be shown as three fixed time boxes. Each single-cycle sprint would require the same background scope definition data as described in the sample above. In each case, the sprint deliverables must then feed into downstream predictive work units as needed. The physical project plan view issue with sprints is to show them as fixed time units as though they were traditional predictive work efforts.

To help conceptualize the sprint logic assume that the goal of a sprint (either type) is to produce a usable set of specifications that are used later. This could be a sub-component that is assembled later into a higher-level product item that could be a predictive fixed assembly task. Note in the example, the sprint is feeding into a traditional predictive work unit.

Another less likely sprint view would be to allow as many sprints as needed to accomplish some desired feature list. Experience suggests that most projects have both time and cost constraints, so the more vanilla iterative plan view would be to show the number of planned sprints as the project plan goal with completion defined based on schedule or cost constraints. Even in the case where the project work is all iterative, the remaining question is focused on the front-end portion which would then drive the work option for the subsequent execution process.

It is important to remind again how the 50/50 work estimating approach nullifies the traditional interpretation of fixed Start and Finish task interpretation. The usage of this task view is so ingrained in the traditional culture that it will be a hard idea to change. However, even in the traditional model, this concept was never an accurate measure of performance given excessively padded task times. In the integrated model, 50/50 work unit calculated dates as shown here do not represent legitimate tracking variables but rather only task sequence. The only schedule-relevant data point occurs when a protection buffer is overrun or consumed faster than planned. An overrun buffer signals that the downstream schedule may be flawed and corrective action is needed to assess the status. With the examples outlined here, one should have sufficient mechanics and understanding to test out the new model processes and combined work execution options.

Chapter 17

Model Background and Implementation

The creation of the integrated project management model described in Chapter 14 was not a planned deliverable. Yet the result simulated a project without planned deliverables. This text evolved from a year-long discussion among the authors cloistered away with Covid protocol with too little to occupy our minds. Discussions regarding project management had been a common theme for years as we meandered through our professional and academic lives. Project failures were observed in a variety of ways and many of the failures come from well-known and avoidable management practices. Over previous periods, our involvement in this industry focused on the production of various items related to topics such as methodology, tools, organizational processes, and the like. It seemed time to leave those low-level views and attempt to describe the broader management process as we saw it. After months of wandering thoughts regarding various project ideas, we began to focus on management approaches such as waterfall, agile, Critical Chain, and others. Our previous experience has provided some exposure to these and other similar methodologies over the year. After several months of gathering background data and thoughts, one early conclusion brought out a broader perspective that in turn produced a modified management view that fit the design concept. This initially looked like an expanded traditional methodology but even then, there were process gaps identified that needed further work. It had some different management characteristics. After matching this structure with the variability of various project types a design skeleton surfaced. Also, an explicit project profile was added to an early decision state to drive the management structure. That was a major conceptual breakthrough. At that point, the expanded decision model view had become more complex regarding linkages and interactions. It was recognized that the strategic

DOI: 10.1201/9781003431091-19

decision level affected tactical decisions, organizational maturity affected the project support processes, and resource issues were universal. Each of these became major management components.

Stage three brought various success factors into consideration and these were matched against the decision block structure. As the decision block process was expanded to the execution level the work options needed to have the flexibility to handle multiple work execution types within the same structure. Three needed work types were identified from classic models—waterfall, agile, and Critical Chain. Pieces of each were described as an optional Lego block process for work unit execution under different situations. A fourth work option type called Modified Scrum was later added to this list to cover predictive work that did not have sufficient requirements definition to utilize the traditional predictive approach. At this point, all of the major process puzzle pieces seemed to be in place and the overall decision block logic was evaluated with different project characteristics.

Stage four focused on cleaning up some of the more abstract ideas that needed more mechanical explanation. Two operational problem areas became more apparent at this point resulting from process changes. First, the concept of 50/50 task estimating was beautiful in that it resolved one of the root problems identified in the waterfall model approach; however, it introduced major negative issues in traditional status tracking and the use of unfamiliar protection buffering logic related to the Critical Chain theory. These two items became the most difficult to justify since they caused a change in the traditional view of status tracking as it took away calendar dates as measures. That is a quantitative task start/finish calendar value. This in turn unearthed the need to replace this with another format. Even more, it led to the recognition that traditional approaches to status tracking already had a major flaw that needed to be recognized as well. This flaw is caused by the padding approach that left task estimates as logically invalid targets. This quagmire of issues took quite a time to sort out, and it still represents the messiest logic part of the new model and is recognized to be a significant acceptance factor.

A second major issue dealt with in stage four was the concept of process flexibility. Risk management is the banner example, but the general point became the recognition that some situations do not follow a consistent decision path or need the same level of rigor. These should be handled by the decision maker with the level of rigor appropriate for that condition based on the project profile. At this discovery point, the puzzle pieces seemed to be sufficiently in place to try to explain them to an outside audience.

The four steps outlined above transpired over several months and were not as clear or discrete as outlined but looking back these steps represent the evolution. Only after a post-review was it possible to see how the fuzzy path evolved. The problem at this point was how to document the model such that it would be logical and positive to a new user. The text outline became the strategy for that. Note that the first 12 chapters describe various background items, which describe the industry, process gaps, or classic models. Starting with Chapter 13 the various decision components and processes were summarized and a layered block structure was introduced. The next chapter proceeded to consolidate the pieces into a formal model structure and key success processes.

It is interesting to note that if this same effort had been undertaken in the traditional project organization, a preliminary deliverable goal and Business Case would have been created for management approval. From that, a planning process would have followed to produce a technical model document. That probably would have resulted in a band-aided version of either a waterfall or agile-type structure, but most surely would not have been the same result. We'll leave the result for the reader to evaluate.

Hopefully, the logical points described in the text have been sufficient to motivate one to try out this new approach. Motivating an organization to change from its current culture or process is a challenge regardless of the topic. This chapter is written to provide some summary thoughts on why and how one might approach the process of utilizing this approach to project delivery. It would seem that the first motivator would have to be a belief that this can improve delivery capability. It is important to see that various decision drivers are embedded in the model that is designed to move the delivery process toward success and away from failure. Concepts related to customized explicit project profiles, modified success parameters, defined success processes, 50/50 task estimating, modified status tracking, and a technique to manage dual work queues should be enough to create some interest in trying out the model.

There will be reviewers of this model who discount its value or approach. Some will reject it because it did not come from a large organization such as PMI or DoD. Successful acceptance of the agile school of thought says that type of sponsor may well not be a good thing. A review of four earlier well-sponsored models that did not achieve great acceptance provides some interesting insights. These are UML, CMM, OPM, and Critical Chain (all mentioned in the text). Each of these models was well documented, had good underlying logic, and was sponsored by a legitimate source. The same can be said for various PMI and DoD documents presented over the years. In some ways, successful support for a new idea seems to be less acceptable when the idea appears to be complex and more successful when it appears to be simple, such as the somewhat vague agile principles exemplify. Agile is the case study for this statement with its loosely defined principles that have allowed the operational mechanics and dialects to evolve naturally at the working level. Case study data correlates the agile iterative approach as being the way to achieve project success, even though the text shows that it has visible gaps in its design. Also, many of the agile principles are not unique and have been essentially described unsuccessfully for use in other traditional models (70% as described in the text). So, the question becomes "what caused agile to be accepted when much of what it was based on was already in the best practices theory?" It is hard to understand why these previous concepts were suddenly recognized as being important. The rest of the agile success story is conjecture but some of it is linked to the core idea of flexible work within the sprint structure. Work decisions seem to be more evident in agile at the lower project team level, and better work focus is a clear productivity value. Because of the advertised success of agile, the iterative concept now has a recognized appeal that needs to be better understood from a total life cycle viewpoint. Iterative concepts are still in transition, so in that regard, the integrated concepts outlined in the new model maybe could have been called "New Agile," even the buzz term "Wagile" might be a good marketing term to fit the new model description. All of these naming options seemed to be a little too presumptuous but the thought was there.

In the case of agile acceptance, one of the factors that seemed to appeal to the industry was the recognized idea that scope could not be defined well in advance and a lot of traditional non-productive formal project management processes were focused on trying to predefine requirements before starting work. Agile attacked that portion of the project with less upfront planning and quicker delivery. The integrated model described in the text accepts that point to a degree but points out that there is more to the project than this subset, and not all work can be looked at as iterative. Hopefully, these points will be accepted as valid and motivate an organization to try out the multiple workstream approach. Just as all projects are not purely predictive, they are also not purely iterative.

The question then moves to how one goes about moving new processes into an existing organizational culture. It is not reasonable to expect instant approval of this or any other organizational change. Some of the new processes may sound good until one starts to define how to do it—i.e., Knowing how to draft project characteristics is one thing, but how does that link to later decisions? This model can be conceptually valid but the implementation portion is the key to long-term usage.

Keys to Acceptance

Logic suggests that the new model contains a valid view of the problem, but it also is evident that it will require various changes in the current approach that will be resisted because it is different. After trying to sell the overall logic of the model the next step is to try it. The recommended approach for that is to use a high-performance team that seems interested in the approach. The target test project should not be a critical initiative as the initial goal is to test out the decision process. However, it is important to select a project that has multiple work execution characteristics. Finally, the project team needs to have startup training on the logic of the model and a good overview of the success recipes and best practices. There are good case studies in the literature outlining change management techniques such as this for introducing new ideas.

In many project management model environments, there are complaints regarding useless work related to topics such as planning, status tracking, and communication. Showing how the integrated model attempts to streamline these topics could be motivational. The one area that has the most potential for acceptance is to cut back on the level of planning detail and move some of that delivery requirements responsibility to the project team and stakeholders. If stakeholders resist being actively involved in this process, one should question whether the project is worth pursuing. It clearly should not be pursued for the amusement of the technical team and stakeholder input as this is a major component of success.

Industry data clearly shows that projects do not have great success. Making such data available to both the project level and management should be a good stimulus to start the conversation about ways to improve the output. It has always been an interesting observation to have an organization say that they execute projects very well but when asked for measurable data they have none. If there is no evaluation of the present, there can be no clear evaluation of a new method. This text perspective has given our best argument as to why previous very logical model approach have not led to desired goal improvement achieved. The starting point for change is to first admit that something needs to be done. It would be best if the change can be stimulated from the lower levels of the organization and even better if done in concert with f top management. Let a working group take the integrated model and decide how they wish to implement something like it. Over time the culture will react to this stimulus. The result cannot be a "no control" model. The delivery process has to meet overall requirements and constraints. Recognize that each of the classic management models has merit but no one of them is a universal solution, nor do they cover the full life cycle. Most importantly, none achieve a reasonable success outcome regardless of the model approach.

Conclusion

By tracking the series of thought processes that led to the integrated model view, one begins to see the iterative project process in action. This exercise was essentially an agile principle with iterative sprints and many years of experience involved with the problem. Would this process have worked if some of the output goals had been defined before starting? Maybe so, but not very well if one looks at the waterfall model structure and this target. And the agile approach would not have assessed the schedule and budget well.

The model described here offers a development view that has the potential to greatly improve deliverable success. Understanding the concept of best practices helps to focus on factors that often lead to lower performance levels.

Chapter 18

Success Recipes

Introduction

The use of a realistic decision-guiding model such as the one described in the text can significantly improve the odds of project delivery success; however, one must also be wary of various associated processes that must be handled properly to support the defined decision steps and achieve the best results. Intermingled with the model architecture and description logic is the concept of best practices to drive the model. The items mentioned collectively drive the project to its conclusion. Failure to properly handle any one of the success practices can cause the project to falter or potentially fail. The text chapters described these practices in various operational environments and the impact was described if not done properly.

This collection of success practices needs to be understood. The description of the integrated model outlined in the text provides a skeleton decision block structure on which these processes are utilized. These blocks do not specifically outline how to execute the process, but they do show the general location, and the text provides guidance on the mechanics, plus offers insights into what needs to be accomplished at various stages. Table 18.1 outlines a selected list of success factors which are deemed to have the most impact on a positive result. Brief general comments are made for each of the items listed in this section.

All of these items were discussed in various text chapters so this section is more of a recap than a full description.

In addition to this abbreviated list, external sources can provide a myriad of other ideas and approaches, so many in fact that it is hard to know which ones are the most important. There are indeed hundreds of other factors that can influence project outcomes. When looking at real-world failure situations, one often finds that the root cause factor should have been obvious. *That is only true in hindsight.* It is important to learn your specific project type and its characteristics regarding problem areas. Other general surveys can also provide insights into problem areas and common failure sources. Chapter 3 made the point that failure factors seemed to repeat across all projects. This should not be the case and it is important to learn from the past as well.

DOI: 10.1201/9781003431091-20

Table 18.1 Success recipe listing

NO.	Topic
1	Achieving Project Success
2	Avoiding Project Failure
3	Use of Templates
4	Project selection process
5	Dealing with Project Risk
6	Defining Requirements
7	Utilizing agile methods
8	Notes on Communications
9	Task Estimating Process
10	WBS and Control Accounts
11	Work chains for speed
12	Speeding up the Project
13	Closing the project

** Note: Each recipe item is referenced by the list number

Items in this success recipe list have been selected from examining process gaps found in traditional projects and related processes gaps uncovered in the research for the text. These are judged to be the most likely items that will be mismanaged and, in turn, negatively impact successful results.

This section is titled *Success Recipes*. Maybe a better title would be *failure variables if not understood*. Each of these topics is attached with a summary outlining why the item is important and the approach that should be taken with it. These have all been mentioned throughout the text, but they are combined here for a quick review. The table of contents list in Table 18.1 has a reference number to help locate that success factor discussion.

Success Recipes

1. Achieving Project Success

The important concept to remember is that success does not come from a fixed set of actions, so the first step in a project is to work with key stakeholders to develop the priority rank order (i.e., schedule, cost, precut function, user satisfaction, etc.).

Second, delivery goals for the project should only be set after a technical analysis of the target requirements. From that analysis of the work required task estimates can be used to derive a plan schedule and budget. Any statement about project deliverables made before this should be looked at as more of a wish or goal statement. Recognize that the project team is the frequent source of blame for failure regardless of the actual source. A significant portion of the project manager's job is explaining to various stakeholders how the project is going and what issues stand in the way.

It is important to understand that project success has many perspectives and variables. The text has refuted the concept that traditional success measurement parameters are always functionality, schedule, and cost. This rejection is based on the wide variety of differences in project characteristics and goals. The text also pointed out that there are varying stakeholder views on this topic, and there is a time dimension to the evaluation (status now or three years from now). To resolve this quagmire of opinions, it is important to discuss this topic with the sponsor, management, and key stakeholders. What often occurs from this is a variety of significantly different goals. Here is an example:

- Sponsor—interested in functionality and secondarily on the completion schedule
- Senior management—major focus on cost
- Key stakeholder—major concern is operational maintainability

As one can see from this example of diverse perspective, the collective goal is to build a widget as specified by all major players while meeting the cost and schedule constraints, in addition to it having "good" maintenance characteristics. This diversity of views must be dealt with early. In many ways, this initial project scope definition is starting to look like an impossible dream and it truly is a stakeholder dream at the initial stage. So, how do you work around this issue? It is very clear that there are several conflicting goals for the project and they are often not compatible. The starting point for resolution is to gather these opinions (wishes/dreams) to assess how bad the diverse views are and then prepare a presentation for this group outlining what a reasonable delivery goal could be. By this point, there should have been some time to assess more background details such as risk, requirements stability, staffing, type of project, and defined constraints. If you are a mathematical type, recognize that this initial goal set is an algebraic equation with no answer. Most importantly, unmanageable situations of this type cannot be left for later if you want to have a chance to deliver a product that is collectively viewed as successful.

A major part of the management domain is to get the key players to understand the complex characteristics of the project. Hopefully, this will bring some flexibility in the defined goals if a reasonable argument can be derived. For example, how hard is the schedule and cost constraint? How fixed is the functionality requirement? What are some specifics of the maintenance requirement? In the beginning, each of the initial external goals is more of a dream than a technical position. The project team is charged with producing the technical output and it does not operate well based on dream logic.

An initial approach for documenting this scenario is to seek an agreement on the rank order for key variables and the hard constraint levels. These have to collectively fit as a delivery set and the subsequent project plan has to fit these. As the project unfolds variability will occur, and this ranking will be important for making proper output tradeoffs. Alternatively, this exercise will help evaluate what items will be compromised based on the rank order. This statement suggests that there is such a thing as prioritized success.

When consensus cannot be reached during the initial goal definition stage, two alternative very hard personal decisions must be made. First, you can state that the goals outlined at this point in your opinion are not viable, and then refuse to proceed and ask for reassignment (probably a career-limiting option). The second and most likely choice is to tactfully say that your research indicates that the delivery goals stated cannot be achieved as stated; however, you understand that these are the desired goals and you will do your best to make them happen. Ideally, you would be given some time to come back with more analysis, but the likely best go forward strategy is to produce a summary document from that session that should be kept for future reference. Later, during status review sessions, the delivery goal status needs to be discussed at least up to the point where they are no longer controversial. Odds are this topic will still be around and the previous summary note will be handy as a discussion reminder. This is the reality in motion!

As this hypothetical example illustrates, some projects have failure built into their initial design regardless of the management model or the skill of the project team. These are called Project Titanic. Here is another scenario example of this. Before the above deliverable goal controversy your boss calls you into his office to offer you a promotion to project manager for the highly visible silver bullet project. There is a significant salary increase for leading this prestigious initiative and the boss wants to announce your promotion today. The project is advertised as a one-year schedule and a five-million-dollar budget. You know nothing else other than rumors of a significant initiative. The key question here is "Do you see failure yet?" In many ways, this is the same position as the one above. You don't know what you are getting into, but it violates your understanding of how project deliverable goals should be formulated. You know that scope should be evaluated first, then work through the mechanics to build a viable plan based on the known details. None of this was done previously. If the project fails, you will be stuck trying to explain why you did not deliver the planned parameters. An even worse example of this can occur when a marketing rep says that his silver bullet item can be installed in two weeks with two people. All of these real-world examples represent Project Titanic.

The success strategy advice for managing the success question for a new project is to deal with this early and try not to be forced into accepting delivery goals from an external group, no matter what their organization level might be (that is easier to write here than live with by the way). There can be a goal statement accepted but be careful even if this tends to become the expected outcome. A misstep at this stage can be career-ending later. Silver bullet projects that are going to solve some major problems seem to be the ones that create Project Titanic most frequently. In the early stages, no one knows exactly what the silver bullet can do just yet. It may be great but often does not do what is claimed.

Now the positive side of this topic. Review Chapter 5 for the general overview of success parameters. Be reminded that non-technical factors such as senior management, stakeholders, host organization support, and basic communications are the more likely root causes of project failure than some of the more logical technical or team issues. Also, these are very hard-to-control items. Given that failure factors seem to hide in all corners, it is necessary to be ready for Murphy's Law to hit and then react quickly before the problem escalates. Failure can come from a lack of aggressive response to these surprise events. Think of these daily perturbations as similar to a small fire. Sit around and watch it for a while and the building burns down. That is a good metaphor to remember regarding not only success but many aspects of project management.

2. Avoiding Project Failure

In many ways failure is the mirror image of success; however, there are additional factors in the project environment that are external to project management but still lead to failure. Some project environments are just not healthy and supportive. And other environmental factors also lead to undesirable outcomes. It is important to identify such factors and mitigate them as much as possible. One common environmental negative factor is a stakeholder that is not supportive of either the basic project goals or who may have alternative views regarding how the project should unfold. Dealing with this type of external conflict is an important exercise outside of just getting the work done.

3. Use of Templates

Typically, one of the least desirable work roles in the management process is the production of various formal documents. Technical team members resist formal documentation, but the use of templates is a recommended aid in satisfying this aspect of the management process and it requires less effort. The time to produce these documents can be dramatically improved through the use of standard templates. A formal library of such templates can help support all aspects of the life cycle with professional-looking documents. Sometimes, the problem in producing a document is finding a way to get started and a template helps with that. This collection of forms can help in standardizing documentation and provide the receiver with a commonly recognized format. Examples of key documents include caned forms in the following areas:

a. Project request, Charter, and Business Case
b. Project scope definition
c. Project plan—schedule and cost
d. Status reporting
e. Risk assessment and tracking formats
f. Team Resources
g. Stakeholder register—A listing of individuals interested in the project
h. Project plan checklist—a reminder of steps

4. Project Selection Process

As the text has stated, "Doing the wrong project right is still the wrong project." Selecting the right project from a large group of proposals is much more complex than it appears on the surface. From a rational view, it is difficult to accurately assess the value of a project proposal at this stage. From a political viewpoint, there is often competing bias and controversy. One way to manage this area is to require more upfront assessment in terms of pilot prototyping and define smaller high-value phases of large projects since this is a major failure category type. Also, grade all project requirements and carefully decide whether the "nice to have" portions should be left in the proposal. Involve all areas of the organization in reviewing the larger initiatives. A formal decision structure should be used in this process area. Chapter 12 provided details on the roles of an EPMO and companion PPM with responsibility options. In the final analysis, there must be active leadership from senior management. No projects should be approved or in work without formal management approval.

Note that industry surveys report a significant number of PMO organizations are in place, but also heavy project failure rates and low life expectancy. Be sensitive to these facts.

5. Dealing with Project Risk

Project risk events can doom a project in many ways, and this aspect of the management process should not be ignored. There are many dimensions to consider regarding project risk. Here are some high spots to consider:

- **Initial Approval**. Always assess the risk related to the technical, resource, and political sides during the initial evaluation. The key question here is to decide what level of review is justified.
- **Planning Phase**. A major decision step involves whether there needs to be a full risk assessment during the planning phase. Research indicates that the current level of risk modeling is immature and may not uncover the full picture. The key question is whether such an intensive process is worth the return—i.e., cost and time versus risk mitigation.

- **Execution Phase**. Regardless of decisions made prior, there is value in establishing a positive risk culture and formally naming risk owners for segments of the project.
- **Risk Focus**. This topic should be discussed openly during status sessions.

**Remember—Project management is risk management; otherwise, everything would just go according to plan.

6. Defining Requirements

The level of scope change during the execution phase is an indicator of requirements quality. Excessive change requests put great stress on the team to stay on track. If this continues, it may be time to call a halt to the project and assess how to proceed, or even cancel the current approach. Chapter 5 outlined the recommended mechanics for handling scope change for both internal and contractual formats.

**Remember—Scope change is the Achilles heel for predictive models and if not handled carefully the project results may be viewed as unsuccessful by some stakeholders regardless of the actual results. Success is measured by one's personal view of the outcome and not necessarily by traditional measures.

7. Utilizing Agile Methods

Agile and its use of sprints have captured the interest of the project community. The text has recognized that there is potential value in iterative task execution and has identified a sprint technique called a Modified Scrum process for predictive tasks. To effectively use the Modified Scrum method, it is recommended that the scope definition be modified to indicate multiple levels of requirements as defined by the MoSCoW idea. This graded approach allows the sprint logic to remain intact and gain the productivity benefits observed with a raw agile model using the looser feature method of requirements specification.

8. Communications

Industry surveys indicate that this deficiency is one of the top reasons for project failure. This gap can occur for the internal team, senior management, or external stakeholders. Spend sufficient time defining the individuals who need to be communicated with, then ask them what format they are most comfortable with and what types of information would be most helpful. This will then define the source, format, and timing for communications to be delivered.

Historically, communication was handled as a "push" process, meaning that fixed formats were regularly sent out in the form of paper reports which were often not read. More recent alternatives came in the form of email and now are moving into internet-based forms. The key to communication is to send it, interpret it, and respond. The delivery technique is only the first step. Every target group has a preferred method. The richest form of communication is face-to-face but it is also the most expensive and therefore used the least.

One contemporary strategy that should be considered is having easy-to-understand data available via a web interface, complete with Google-like search capability. This format is called a "Pull" model where the recipient can browse for whatever level of detail they desire. This is also an area where the host organization can be very supportive with a standard access model that is easy to import project data into.

Data collection and administration can be a very expensive activity for the project team, but it is important to recognize that the project is being influenced by many other players and communication is necessary to keep all players in the loop and not have a hidden misunderstanding.

9. Task Estimating Process

The revised approach to task estimating described in the text is considered to be one of the vital methods to achieve success. Review the material in Chapter 15 regarding the underlying logic of this non-traditional approach.

The 50/50 estimating approach is a necessary strategy to activate the project but it brings with it the increased complexity of status tracking that must be understood by external reviewers. Compared to traditional monolithic models with a single work queue this new approach can be viewed as complex. An example of this was shown in Chapter 14 with a hybrid dual work structure using traditional predictive execution and Modified Scrum options occurring in parallel.

Estimating Issues When Using Buffers and Reserves

Buffers described in the integrated model are justified based on the 50/50 task estimating logic which is borrowed from the Critical Chain theory outlined in Chapter 9. Specific buffer mechanics are described in Chapters 14 and 15. The use of buffers is necessary to protect key completion dates but this mechanic destroys traditional status-tracking views. A clear understanding of this mechanic is required as it will necessitate significant changes in reporting project status.

The sample project plan shown in Figure 18.1 illustrates a modified approach to buffering. In this example, there are four buffers shown (IDs 7, 14, 18, and 23). The first three buffers are attached to major project groups, while ID 23 is inserted to protect the entire project. There are many reasons why buffers might be utilized in this fragmented format. The most realistic logic is to protect that segment based on the type of work options employed. In this example, status tracking by phase is enhanced by using multiple phase-level buffers and a smaller project completion

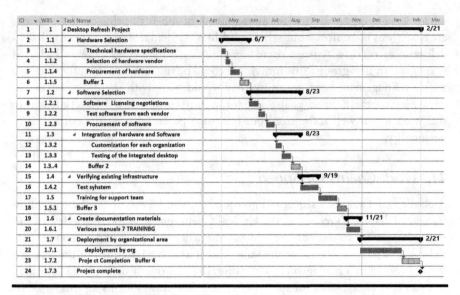

ID	WBS	Task Name	Apr	May	Jun	Jul	Aug	Sep	Oct	Nov	Dec	Jan	Feb	Mar
1	1	⊿ Desktop Refresh Project												2/21
2	1.1	⊿ Hardware Selection		6/7										
3	1.1.1	Ttechnical hardware specifications												
4	1.1.2	Selection of hardware vendor												
5	1.1.4	Procurement of hardware												
6	1.1.5	Buffer 1												
7	1.2	⊿ Software Selection					8/23							
8	1.2.1	Software Licensing negotiations												
9	1.2.2	Test software from each vendor												
10	1.2.3	Procurement of software												
11	1.3	⊿ Integration of hardware and Software					8/23							
12	1.3.2	Customization for each organization												
13	1.3.3	Testing of the integrated desktop												
14	1.3..4	Buffer 2												
15	1.4	⊿ Verifying existing infrastructure						9/19						
16	1.4.2	Test syhstem												
17	1.5	Training for support team												
18	1.5.1	Buffer 3												
19	1.6	⊿ Create documentation materials								11/21				
20	1.6.1	Various manuals 7 TRAININBG												
21	1.7	⊿ Deployment by organizational area												2/21
22	1.7.1	deplolyment by org												
23	1.7.2	Proje ct Completion Buffer 4												
24	1.7.3	Project complete												

Figure 18.1   Phase Buffering

buffer. This is not a typical use of the concept but illustrates the flexibility intended in the model.

10. WBS and Control Accounts

The WBS format represents one of the most useful and flexible project artifacts. Boxes in the WBS represent both individual and collections of work to be performed. The integrated model describes various roles for this hierarchical structure. At the highest level, it can be used to show major phase partitions of the project delivery strategy. Middle levels of the structure can highlight major subsystems or work groups. The lower levels can describe work packaging decisions. By adding task estimates and work sequencing data to this view, an initial project plan can be produced using standard software. Initial project planning can be enhanced with the use of WBSs to outline major items.

In addition to the visual value of this utility, schedule and cost data can be embedded for use in project communications. WBS IDs become frequent communication labels for various processes.

A companion to the WBS is the associated WBS Dictionary. This formal data store is recommended as a source of project data. Used properly, it can contain the history of the project and all relevant data (or links to relevant data). One contemporary use of the WBS that was not mentioned in the theory portions of the text is the concept of a Control Account (CA) and a Control Account Manager (CAM). A CA can be a single WBS box or a collection of boxes. This designation is useful for recording actual schedule and budget values. In cases where there is organizational standardization of the WBS, this option provides cross-analysis of projects, but when used in a particular project, it formalizes a delegation structure. Assume some box grouping is defined as a CA and a CAM is assigned to that group. This individual could be the technical lead for that area with a management bent, meaning they would work with the project manager on achieving the defined goals. The formalization of a CAM function within the WBS is not well recognized but has great potential with the correct team personality.

11. Work Chains for Speed

A classic task chain is a sequential group of tasks to be executed. These are illustrated in Chapter 10 related to the Critical Chain model. A modified view of this time compression strategy can be structured as a defined group of even non-sequential tasks to be executed using the same principles. This is defined in the text as a *work chain*. From a management viewpoint, this is formally defined as a collection of tasks that are targeted for special attention. The reason for using this could be time compression, risk monitoring, or any other reason, although the most common reason for this is to reduce or recover cycle times. Think of a work chain as a task group focal point for work execution.

Work chains can be part of the initial plan, or more likely used as an emergency reaction to an event that requires a higher level of control and execution. Regardless of the reason, multiple project segments can be isolated this way for special treatment. There are multiple segments in a project where some grouping of tasks may be best handled using this approach when it may not be feasible to execute the whole project this way. Think of the time compression work chain being executed using Critical Chain principles. In the road construction example described in the text, the congestion continued unresolved for months when the creation of a small work chain could have cleared the task problem area in a short period. This represents a real example of management getting lost in the work plan and forgetting about what the project goal is (i.e., avoiding congestion). In the meantime, workers are toiling away elsewhere to pave a portion of the road that is not causing undue congestion. The concept of work chains does not appear in the literature as a management idea, but given the value found in the Critical Chain theory, this approach has a place in the management tool kit.

12. Speeding Up the Project

One of the most frequent questions a project manager gets is "Why is this taking so long." Several documented techniques have the potential to speed up the life cycle. Many of these can be effective in varying degrees and some may actually produce the opposite result. In this recipe, we will review the following strategies that are designed to cut project time:

1. Work chains—covered in recipe #11
2. Critical chain—covered in Chapter 10
3. Lean analysis
4. MoSCoW technique
5. Fast Tracking
6. Crashing
7. Reduce documentation
8. Curtail changes
9. Cutting out "soft" steps
10. Working overtime

Each of these items deserves a brief comment, but they all need to be understood in the working tool kit of the project manager.

Work Chains. Review this topic in recipe #11. It represents a workable management technique with a proven process.

Critical Chain. This topic was covered in some detail in Chapter 10. It is considered to be a valid technique but requires significant setup and learning time.

Lean Analysis. The meaning of this term needs a clearer definition. If "Lean" means "less," that can leave a gap in a required deliverable process that will cause trouble later. On the other hand, if it means "Lean" in a more technical sense, that implies the use of various techniques such as graded requirement ideas as defined by the MoSCoW idea, Creative concepts of this nature offer increased flexibility in defining and producing various work processes. This is a legitimate strategy to cut scope as needed to improve cycle time. In the traditional project environment, this is called the functionality/Time/Cost tradeoff. The tradeoff goal here is Time versus some modified process. Performed properly, this is a valid technique.

MoSCoW technique (Must Have, Should Have, Could Have, and Want).— This approach to defining requirements is occasionally mentioned in the literature but seldom done. The ability to view requirements in this manner has significant management value for work execution. This is one of the recommended success methods outlined in the integrated model. A graded requirement allows greater future tradeoff flexibility in the work process.

Fast Tracking. This technique is defined in the traditional project model, but it also applies to any process situation. If two sequential tasks can be converted to run in parallel, the cycle time would be decreased by the amount of overlap. This process becomes tricky with the output of the predecessor task requiring some level of finished work before the predecessor can begin.

Crashing. This is a traditional companion method to Fast Tracking. In this process, task duration is cut by adding resources to it. If the task is estimated for one resource and two are assigned, the theory says the cycle time should be reduced. One must be careful in examining the degree to which adding additional resources is valid.

Reduce Documentation. This is another form of scope reduction. The value of formal documentation has long been discussed since it is viewed negatively as less worthy in the overall scheme. Few professionals would choose this work activity. One way to reduce the effort to create required documents is to use templates (see recipe #3) and use edited archived project files for similar documents that could be plagiarized (legally of course). In both of these cases, the effort required is reduced. Omitting required documents is not a recommended option and should be done only after careful review.

Curtail Changes. There would be a significant saving in cycle time with the elimination of changes, but this is a dangerous option. If the change request process is operated properly, all approved items have been judged to improve the final product. So, stopping that flow by definition curtails the associated improvements. There is one change strategy that might be viable. That is, to make only absolutely needed changes and move all others to a future project cycle.

Cutting Out "Soft" Steps. The term "soft" in this case means certain tasks that don't directly add to the core product. Two typical examples of this would be to cut out a risk assessment and don't perform a full slate of testing. Both of these options decrease project scope and technically would save time unless a risk event that could have been avoided did occur, or the negative event that caused the project to fail that could been uncovered by the missing test before product release. Cutting tasks now may have adverse effects later. To offer one more example type to watch out for, "Let's cut out the monthly birthday celebrations because it wastes team productive time." Watch out for the flawed logic of cutting soft events.

Work Overtime. If there is a common method for decreasing project cycle time, this is it! Salaried employees generally do not get paid overtime, so this practice creates more project resource time for free. The reality of projects says that some overtime is often needed, but this should not be the extra resource solution. Extensive overtime erodes morale and productivity. Don't ignore this potentially negative culture. One organization had a uniquely positive way of handling overtime. When this occurred, the project manager would be sure to publicly thank the team members for their efforts and would frequently give them time off, Gift cards, or even $100 bills. When telling this story, it is surprising to have the audience say that this seems unnecessary, particularly the money option. However, it a professional has worked extra for a week on a special task does that not seem worthy of $100? Regardless of the management approach, beware of requiring excessive overtime as it is a frequently overused used option.

13. Closing the Project

This topic was not described in the model description and therefore is not an obvious success process. For that reason, more detail is needed here. Improper closure has a potential impact on future events. There are at least 13 processes that need to be executed before considering the project to be completed. These are:

a. *Obtain client acceptance*: The client formally verifies and accepts the project deliverable, and this event is formally documented.

b. *Transition deliverables to owner*: The team formally hands off the project deliverables to the new owner. This includes possession of the item and the ability to support it.

c. *Closeout contract obligations*: The project team will coordinate with procurement personnel to document the status of all contractual relationships.

d. *Update the organization's central information repository*: This activity involves documenting project records and deliverables as a formal archive for the organization.

e. *Document final project financials*: This includes a budget status summary and variance analysis.

f. *Close various accounts and charge codes*: This activity involves the process of closing team member accounts and codes related to financials, infrastructure, and security.

g. *Update resource schedules*: Work to ensure that team members have appropriate job opportunities following the closure.

h. *Conduct performance evaluations*: The PM must ensure that appropriate performance feedback is performed and documented for all team members.

i. *Update team resumes*: The team members should update their resumes to reflect the project experience.

j. *Publicize market project accomplishments*: Formally recognize team member accomplishments and overall project positive experiences.

k. *Review project performance with clients*: This process evaluates whether the team achieved the desired goal from stakeholder viewpoints.

l. *Celebrate*: From a team morale standpoint, it is important to find something to celebrate after a project concludes. Try to leave the project team feeling positive about their experience.

m. *Capture lessons learned*: Documenting the project team's experience-related activity enables future projects to profit from both good experiences and mistakes made.

Four of these items are less obvious and deserve additional discussion.

Contractual Closing Example

Leaving contractual status unresolved is a lawsuit waiting to happen. On the surface, this may not seem likely because the original terms are clear but during the execution phase items could have been sent back, failed in testing, performance incentives, lost items, etc. One side thinks the item is clear, and the other disagrees. This scenario represents a subtlety that does not seem important, but it is likely the most dangerous area to omit in project closing. Recognize that after a project ends, the team is often dispersed and, in some cases, leaves the organization. The basic goal of contract closing is to match actual events to documented contractual terms along with any addendums. This requires matching deliverables with fees, plus any other terms. Once the audit is complete, a formal meeting with the vendor should be held to review the status. That status should be summarized in writing and signed by both parties. Every effort should be made to leave the contract's agreed-upon status as clean as possible. Recognize that any conflict item is a potential lawsuit, so deal with this activity accordingly. It can be difficult to reconstruct all of the machinations that have occurred throughout the life cycle. Once the project team has dispersed, the internal knowledge is also gone, so written documentation will be important.

Equipment

In each of these areas, the concern is to document the current status regarding what needs to be done with it. The following scenario examples will help justify this as a required activity:

a. Suppose you loaned equipment to a vendor to help with the contract. The vendor now thinks you gave it to them. This is the time to resolve that issue.
b. You have been under the impression that the contract was finished when all of the tasks were completed and the customer accepted the product. A senior executive made a verbal commitment to provide one year of training to the new users (yes, that happens). Now you have an unplanned staffing requirement.
c. Similar to the item above, the contract specifies customer warranty support for the product for one year and your team is the only group capable, but this was not formally included in the approved plan. Another staffing issue to resolve.

Closing Celebration

This is one of the least understood management processes. The goal at this stage is to help the team leave the project with the belief that their efforts were appreciated and worthwhile regardless of the state of completion. In many cases, the project has been difficult and may even have been terminated early. The team needs to feel like they accomplished something regardless of the circumstance. Later, when team members get back together to share memories, you will hear a sentence starting with "Do you remember when?" Team members will view the project as a learning experience and one that is not so bad in hindsight. If all has gone well, this is easy but try to find something positive to share and congratulate the team members regardless of the situation. A team formal closing celebration can be a simple pizza lunch, or an expensive couples evening out with full regalia. This get-together should be planned to generate a positive reminder that the project had some good aspects to it. Of course, a monetary performance reward bonus would be even better and should be considered. The management goal is to make the closing a positive event. If the project manager has done his or her job, the team will be able to look back and see that their professional expertise was improved.

Lessons Learned

One area of project management that is viewed with increased interest is a formal lesson-learned process. This statement is true for the current project but maybe even more valuable for future projects. Mature organizations have processes in place for internal evaluation of status and associated corrective action in areas that need improvement. The lessons-learned process serves a major part of that role. From an operational viewpoint, many of the project artifacts have positive sharing potential (i.e., Business Case, Charter, WBS, task estimates, plans, and templates of various kinds). Making mistakes from one project to the next is a common occurrence to be avoided.

Index

Printed in the United States
by Baker & Taylor Publisher Services